Adaptive Governance in Carbon Farming Policies

Nooshin Torabi

Adaptive Governance in Carbon Farming Policies

palgrave
macmillan

Nooshin Torabi
School of Global, Urban and Social
  Studies
Royal Melbourne Institute
  of Technology
Melbourne, VIC, Australia

ISBN 978-3-319-97495-8   ISBN 978-3-319-97496-5   (eBook)
https://doi.org/10.1007/978-3-319-97496-5

Library of Congress Control Number: 2018949620

Cover illustration: © Melisa Hasan

This Palgrave Pivot imprint is published by the registered company Springer Nature
Switzerland AG
The registered company address is: Gewerbestrasse 11, 6330 Cham, Switzerland

To my twin boys (Radmehr and Radin) for their constant engagement in this book's progress.

FOREWORD

Effective environmental governance is challenging, encompassing intersecting biophysical and social processes with complex and dynamic interactions. Complex and dynamic processes, uncertain understandings, multiple threats, multiple jurisdictions and scales, multiple stakeholders and multiple perspectives (which intersect in ways that are mutually informing) are recurring elements encountered in our search for socially just and ecologically sustainable forms of development.

Effective environmental governance clearly represents a considerable challenge, whatever mode of governance is preferred—rules-based, competition-based, networked, or hybrid. Further, while hybrid and interactive modes of governance offer flexibility, there is nothing in them which makes them inherently more sustainable: for example, public–private partnerships may drive ecological degradation. "Nature" simply does not conform to the dictates of the administrative mind, market logic, vagaries of social relations or any particular combination of different modes of governance.

Given the nature and magnitude of the social and environmental challenges facing humanity, urgent attention needs to be directed to articulating ideas about development, and enacting forms of governance, that promote social justice and ecological sustainability: incremental responses will not work. What is required is a "whole of governance" orientation, where people are individually and collectively engaged in reimagining what it means to live with other species on planet earth. This is an ongoing and multifaceted endeavour—there are no silver bullets and there is no one single source of authority: science and technology will not save the day.

Within this context, carbon framing is one approach that is promoted as a means of grappling with some of the myriad challenges we face in the Anthropocene. Informed by ideas about adaptive governance and knowledge exchange, Dr. Torabi investigates the potential of carbon farming initiatives in Australia as a means for tackling some of the challenges associated with climate change. Based on detailed empirical work, various elements involved in carbon farming are discussed within the context of the literature on adaptive governance and suggestions for the design of programs discussed.

Melbourne, VIC, Australia   Dr Brian Coffey
May 2018

ACKNOWLEDGEMENTS

I would like to acknowledge landholders, policymakers and academics who kindly participated in this research. I was able to undertake it only with their willingness to share their knowledge and experiences.

CONTENTS

Contents

LIST OF TABLES

Introduction: Biodiverse Carbon Plantings as a Tool for Mitigating and Adapting to Climate Change

Abstract This chapter offers an extensive overview of climate change mitigation and adaptation strategies. The Australian carbon farming policy as one of the means to tackle the wicked problem of climate change has been discussed in detail. This chapter also explores the role of landholders as change agents in the uptake of carbon farming policies.

Keywords Climate change · Carbon farming · Carbon policy
Landholders · Mitigation · Adaptation

CLIMATE CHANGE

The issue of climate change has received substantial public and scholarly attention and is now a strategic part of the global economic and ecological consideration (Capoor and Ambrosi 2008). As the human-induced change in climate continues (IPCC 2014, p. 118) (IPCC and Cambridge University Press 2007), impacts on ecosystems and society will become increasingly problematic, particularly if greenhouse gas emissions (GHGs) are not halted (Wise et al. 2014). The wicked problem of climate change conveys an urgency for decision-oriented studies seeking to specify immediate mitigation and/or adaptation actions (Wise et al. 2014). Mitigation refers to those aiming to reduce greenhouse gas (GHG) emissions either at their sources or by using sinks to sequester the emissions (Cole et al. 1997). Adaptation actions are "adjustments in

© The Author(s) 2019
N. Torabi, *Adaptive Governance in Carbon Farming Policies*,
https://doi.org/10.1007/978-3-319-97496-5_1

natural or human systems in response to actual or expected climatic stimuli or their effects, which moderates harm or exploits beneficial opportunities" (IPCC 2011, p. 72).

Tree planting is considered an important means to pursue both mitigation and adaptation to climate change. Hence, incentives to accelerate tree planting have become an important component of climate change policy. This is because of the ability of trees to sequester carbon more than other terrestrial ecosystems (Gibbs et al. 2007). Thirty per cent of the Earth is covered by trees, and 45% of terrestrial carbon is stored in forests (NASA Earth Observatory 2012). Forests contribute to 50% of net primary production in the world (Sabine et al. 2004).

Mitigation Outlook at the Global Level

The role of market-based mechanisms to encourage GHG emissions reduction was highlighted in the Kyoto Protocol (Niesten et al. 2002). Endorsed by 182 countries, the Protocol is a legal agreement with the commitment from developed countries to reduce their GHG emissions 5.2% below 1990 levels between 2008 and 2012 (Bayon et al. 2007). The most significant outcome of this protocol to mitigate global warming is the establishment of a carbon market (Capoor and Ambrosi 2008). However, the commitment period of the Kyoto Protocol ended in 2013 and the Durban Climate Change Conference 2011 focused on finding various methods of emission reduction based on international negotiation (UNFCCC 2011). It is regarded as a "roadmap" towards a global agreement (Hill 2011). "… It enables a second commitment period to start on 1 January 2013 as part of a transition to a wider single global and comprehensive legally-binding agreement" (Council of the European Union 2012, p. 3).

The global carbon market falls into two categories: regulatory and voluntary markets. Voluntary markets focus on project-based offsets (forestry, methane destruction), whereas regulatory markets operate under the cap-and-trade and baseline and credit schemes (Bayon et al. 2007). Some examples of regulatory markets are the European Union Emission Trading Scheme, the Chicago Climate Index, the New South Wales Greenhouse Gas Reduction Scheme (GGAS) and the Alberta Offset System (Bayon et al. 2007; Brohé et al. 2009). The global carbon market was valued at US$142 billion in 2010, including a US$424 million share of the voluntary market (Peters-Stanley et al. 2011; Linacre et al. 2011). REDD (Reducing Emissions from Deforestation and Forest Degradation) projects constituted 29% of the voluntary carbon market (Peters-Stanley et al. 2011). REDD is a United Nations programme to reduce emissions from forested plots by providing incentives and valuing the carbon sequestered in trees in developing countries (FAO UNDP UNEP 2008). However, critiques of the programme demand for an improved governance system to regulate and protect forests (Adelman 2015).

Adaptation

Historically, climate mitigation had been the principal mechanism considered by researchers and policymakers to tackle climate change (Heller and Zavaleta 2009). However, climate change adaptation is gaining more attention as reliance on mitigation methods is manifestly inadequate in the face of current impacts and climate projections in the immediate future (Wise et al. 2014). Adaptation aims to reduce the climate change vulnerability (Spittlehouse and Stewart 2003). Adaptation occurs as a dynamic practice in the societies, and it could help to provide an economic improvement over cases where no adaptation takes place (Adger 2013). These actions happen at different social and institutional levels and on socio-economic and political scales.

Carbon plantings are considered to be both mitigation and adaptation strategies. The practice helps to mitigate GHGs and reduces the vulnerability of ecological and social systems to climate change (Van Noordwijk et al. 2011). However, it is important to be aware of the uncertainty involved in undertaking such plantations as climate is changing (Hulme 2005) at the same time the socio-ecological and the political contexts are (Adger et al. 2005). Actions that are taken now need to take into account climate change projections to make sure different species will survive under new climate projection scenarios (Hulme 2005). It is also essential to understand the context, management requirements and institutional and societal necessities of successful outcomes (Spittlehouse and Stewart 2003). Furthermore, various drivers for environmental degradation require thorough investigation in the socio-ecological systems (Commonwealth Scientific and Industrial Research Organisation [CSIRO] 2003).

State of GHG Emissions in Australia

Australia is the world's 15th highest GHG emissions polluter and contributes 1.3% of global emissions (Commonwealth of Australia 2014). There will be 421 Mt of CO_2 emissions from the current levels by 2020, and abatement to reach the current target of at least 5% reduction (of the 2000 level) will be 131 $MtCO_2$-e (Commonwealth of Australia 2013). As a hot and dry continent, Australia will be much affected by climate change with more frequent droughts and fire (Garnaut 2008; Buys et al. 2011). Hence, immediate actions to mitigate these impacts are necessary (Wise et al. 2014).

Biodiversity

… Most of the potential risks and surprises affecting biodiversity also present opportunities if Australians think strategically, anticipate, prepare and act. (Hatton et al. 2011, p. 39)

In Australia, biodiversity is in a parlous situation and biodiversity loss is among the most important ecological issues (Hatton et al. 2011; Vanclay and Lawrence 1995). The State of Environment Report 2011 (Hatton et al. 2011) concluded that human activities such as land clearance in addition to population growth are responsible for the situation. Currently, public conservation areas encompass nine per cent of Australia but are not considered adequate to conserve biodiversity given their size and the ecological systems they represent (Cowell and Williams 2006). Therefore, biodiversity conservation on private land (two-thirds of Australia) requires more attention and the participation of landholders is essential (Stephens 2001).

Policy Setting

Carbon Policy

The carbon market in Australia has been voluntary (excluding the NSW GGAS), and there was no national cap-and-trade mechanism. However, carbon offset providers have been offering a range of products around Australia (carbon plantings, renewable energy). A carbon tax was approved by the Australian parliament to affect the top 500 polluters, with an initial carbon price of $23 a tonne in July 2012, moving to an emission trading scheme in 2015 (Australian Government 2011). The carbon tax was repealed as of July 2015, and a Direct Action Plan is proposed by the new government as its climate action policy.

As a central part of Direct Action Plan, the Emission Reduction Fund (ERF) White Paper was released in April 2014, aiming to serve both the economy and the environment (Commonwealth of Australia 2014). It works based on the reverse auction mechanism for businesses and communities to sell emission reduction projects to the government, meeting Australia's five per cent target below 2000 levels by 2020, with the aim of a total of $1.55 billion allocated funding (Commonwealth of Australia 2014). Approved methods under the ERF include land sector methods such as biodiverse carbon plantings and reforestation.

The Carbon Farming Initiative (CFI) was enacted in the Australian Federal Parliament in August 2011 aiming to reduce emissions and establish tradable carbon credits (Australian Carbon Credit Units, or ACCUs) through enhanced land management practices (Australian Government 2011). CFI aims to achieve 4 $MtCO_2$-e of abatement from activities such as deforestation and reforestation (Commonwealth of Australia 2012). Since the start of the scheme, 14,591,415 ACCUs have been issued (Australian Government 2013).

Biodiverse carbon planting is a key private land conservation practice that needs active stakeholders' involvement to deliver successful policy design and implementation. In addition to storing carbon, tree planting has the potential to preserve vital ecological processes and provide suitable habitats for wildlife (Bauhus et al. 2010; Campos et al. 2005; Carswell and Burrows 2006). Biodiverse plantations will potentially increase the availability of resources for native animals, function as seedling banks and enhance the resilience of the ecosystem against climate change and pest invasion (Crossman et al. 2011; Pearce 2005). Such plantations can be incorporated into existing farming systems through windbreaks, riparian zones and native woodland plantations (Sabto and Porteous 2011).

Change Agent's Role

The increasing public attention on climate change is expected to generate a market for greater investments in bio-sequestration projects and the restoration of biodiversity (Bekessy and Wintle 2008). To encourage participation in biodiverse plantings for carbon sequestration, private landholders and investors should be able to take advantage of both ecosystem service markets (the carbon market and biodiversity market) on the one piece of land (Bekessy and Wintle 2008). The global market for biodiversity is at least US$2.4–4.0 billion annually, but 80% of current projects are too vague to evaluate their market size (Madsen et al. 2011). These two global environmental service markets have the potential to help private landholders generate income while benefiting both climate change abatement and biodiversity management.

Many private land conservation programmes fail to achieve sufficient landholder uptake (Comerford 2014). In the case of biodiverse carbon plantings, a better understanding of landholders' sociocultural drivers (non-market values) and how these relate to programme design and financial incentives could assist with delivering a scheme that could better achieve biodiversity conservation and carbon sequestration objectives. Such considerations could lead to higher participation rates and help to close the policy-implementation gap (Kragt et al. 2014), hence improving the projected outcomes of policies. Schemes that consider landholders as change agents in programme design and implementation are likely to be more successful (Blackmore and Doole 2013).

Market approaches and payments for ecosystem services aim to increase biodiversity conservation by providing financial incentives to landholders. Biodiverse carbon planting is one of these market mechanisms that have the potential to stimulate investment in biodiversity conservation alongside carbon sequestration. However, there has been some criticism of the lack of ecological considerations including proper monitoring of biodiversity outcomes (Burns and Lindenmayer 2012). Furthermore, social and cultural factors have the ability to influence these market approaches to biodiversity management, yet they are often overlooked in the design of programmes (Bekessy and Cooke 2011). Indeed, Walker et al. (2009) questioned the ability of market-based schemes to achieve preferred biodiversity outcomes in dealing with complex socio-ecological systems. Failing to appropriately address the sociocultural drivers will reduce investment effectiveness (Bekessy and Cooke 2011).

In addition to social, environmental and cultural drivers for participation, the attractiveness of a programme to landholders could influence participation rates. McCann (2013) states that there is a lack of research in areas related to the design of environmental policies. The success and effectiveness of policies that aim to provide payments for ecosystem services (i.e. biodiverse carbon planting) depend on the design of the programme offered to landholders (Engel et al. 2008). When a carbon farming programme is offered to landholders, its characteristics could influence the likelihood of landholders participating. For example, a more flexible programme could fit more easily with existing land management approaches and hence may be more appealing (Blackmore and Doole 2013).

Financial incentives are traditionally considered a strong motivation for landholders to participate in private land conservation schemes (Rode et al. 2015). However, recent research suggests that such incentives may have minimum impact on landholders' decisions to participate in carbon planting programmes (Kragt et al. 2014; Blackmore and Doole 2013). Although monetary incentives could offset transaction costs including establishment and ongoing monitoring expenses (Cacho et al. 2013; Cacho and Lipper 2007; Bigsby 2009), they are not necessarily driving landholders' willingness to participate. Considering incentives to be the main motivation could "crowd out" the primary drivers for participation (Luck et al. 2012; Rode et al. 2015). This issue highlights the potential for complex sociocultural factors to influence programme uptake and implementation success.

Although biodiverse carbon planting is an environmental practice, it is important to consider its social and economic aspects. A successful policy design needs to consider social–cultural factors in landholders' acceptance and participation in a new scheme. In this book, I investigate the various elements that are involved in landholders' decisions to participate in such practices including their sociocultural drivers, the characteristics of a programme and financial incentives offered. I also elaborate the role of adaptive governance in theory and practice in the success of carbon farming programmes.

SIGNIFICANCE

Significant outcomes of this research will include: (i) improving the decision-making capacity of government and policymakers involved in managing carbon and biodiversity markets; (ii) helping to choose a particular course of action to engage landholders in more effective land conservation; and (iii) generating enhanced biodiversity outcomes by considering sociocultural drivers. At the national level, this research will explicitly tie into two Federal Government National Research Priority Goals, namely "Australia's Biodiversity Conservation Strategy 2010–2030", which aims to double the value of markets for ecosystem services by 2015, and "An Environmentally Sustainable Australia", the national research goal of sustainable use of Australia's biodiversity (Department of Sustainability, Environment, Water,

Population and Communities 2010). The book will broadly contribute to our knowledge of the extent to which biodiverse plantings for carbon sequestration are influenced by multiple stakeholders' viewpoint. It also investigates how progressed adaptive governance is in communicating challenges and opportunities facing design and implementation of carbon farming from science, policy and community perspectives.

OVERVIEW OF THE BOOK

This book comprises five chapters. Following this introductory chapter, in Chapter 2, I present a review of the rich literature on adaptive governance. This chapter also focuses on the urgency to hear multiple stakeholders' voices in design and implementation of conservation policies. Chapter 3 addresses the method and methodology of this study, followed by outlining the sociocultural factors driving landholders' participation in biodiverse carbon plantings. This is based on a survey of landholders and interview results. Surveys reveal demographic data while interviews provide a deep understanding of various aspects of adoption. In Chapter 4, similarities and differences in stakeholders' voices will be presented. It aims to compare results from the interviews with landholders and other stakeholders. It also focuses on the theory–practice gap in adaptive governance. Chapter 5 presents a broad discussion of findings, including recommendations for policy development. It concludes the book, seeking to broaden findings to environmental management in other contexts.

REFERENCES

Adelman, S. (2015). Tropical Forests and Climate Change: A Critique of Green Governmentality. *International Journal of Law in Context*, *11*(2), 195–212. Available at: http://www.journals.cambridge.org/abstract_S1744552315000075. Accessed January 25, 2016.

Adger, N. W., Arnell, N. W., & Tompkins, E. L. (2005). Successful Adaptation to Climate Change Across Scales. *Global Environmental Change*, *15*(2), 77–86. Available at: http://linkinghub.elsevier.com/retrieve/pii/S0959378004000901. Accessed April 28, 2014.

Adger, W. N. (2013). Social Capital, Collective Action, and Adaptation to Climate Change. *Economic Geography*, *79*(4), 387–404.

Australian Government. (2011). *Carbon Price*. Available at: http://www.cleanenergyfuture.gov.au/clean-energy-future/carbon-price/.

Australian Government. (2013). *About the CFI*. Canberra. Available at: http://www.cleanenergyregulator.gov.au/Carbon-Farming-Initiative/Reports-and-publications/Pages/ACCUs-issued-in-Q2-of-2014.aspx.

Bauhus, J., van der Meer, P., & Kanninen, M. (2010). *Ecosystem Goods and Services from Plantation Forests*. London: Earthscan. Available at: http://rmit.eblib.com.au/patron/FullRecord.aspx?p=585453.

Bayon, R., Hawn, A., & Hamilton, K. (2007). *Voluntary Carbon Markets: An International Business Guide to What They Are and How They Work Second*. London: Earthscan.

Bekessy, S. A., & Cooke, B. (2011). Social and Cultural Drivers Behind the Success of Payment for Ecosystem Services (PES). In D. Ottaviani & E. Scialabba (Eds.), *Payments for Ecosystem Services and Food Security*. Rome: Food and Agriculture Organization of the United Nations.

Bekessy, S. A., & Wintle, B. A. (2008). Using Carbon Investment to Grow the Biodiversity Bank. *Conservation Biology, 22*(3), 510–513. Available at: https://ezp.lib.unimelb.edu.au/login?url=http://search.ebscohost.com/login.aspx?direct=true&db=eih&AN=32549766&site=ehost-live.

Bigsby, H. (2009). Carbon Banking: Creating Flexibility for Forest Owners. *Forest Ecology and Management, 257*(1), 378–383. Available at: http://linkinghub.elsevier.com/retrieve/pii/S0378112708006920. Accessed May 12, 2014.

Blackmore, L., & Doole, G. J. (2013). Drivers of Landholder Participation in Tender Programs for Australian Biodiversity Conservation. *Environmental Science & Policy, 33*, 143–153. Available at: http://linkinghub.elsevier.com/retrieve/pii/S1462901113001226. Accessed November 11, 2013.

Brohé, A., Eyre, N., & Howarth, N. (2009). *Carbon Markets: An International Business Guide*. London; Sterling, VA: Earthscan.

Burns, E., & Lindenmayer, D. (2012, February 1). *The Biodiversity Fund— Another Missed Opportunity?* Available at: http://theconversation.edu.au/the-biodiversity-fund-another-missed-opportunity-4889.

Buys, L., Miller, E., & Megen, K. (2011). Conceptualising Climate Change in Rural Australia: Community Perceptions, Attitudes and (In)Actions. *Regional Environmental Change, 12*(1), 237–248.

Cacho, O., & Lipper, L. (2007). Abatement and Transaction Costs of Carbon-Sink Projects Involving Smallholders. *Fondazione Eni Enrico Mattei, Nota di Lavoro CCMP, 27*(1).

Cacho, O. J., Lipper, L., & Moss, J. (2013). Transaction Costs of Carbon Offset Projects: A Comparative Study. *Ecological Economics, 88*, 232–243. Available at: http://linkinghub.elsevier.com/retrieve/pii/S0921800912004910. Accessed February 4, 2013.

Campos, J. J., Alpizar, F., Louman, B., & Parrrotta, J., & Porras, I. T. (2005). An integrated approach to forest ecosystem services. In Mery, G., Alfaro, R., Kanninen, M. and Lovobikov, M. (Eds.), *Forests in the Global Balance— Changing Paradigms. IUFRO World Series Volume 17*. Vienna, Austria: International Union of Forest Research Organization.

Capoor, K., & Ambrosi, P. (2008). *State and Trends of the Carbon Market 2008*. Washington, DC: World Bank. Available at: http://go.worldbank.org/4BDX5VOLH0.

Carswell, F., & Burrows, L. (2006). Could Biodiversity Add Value to New Zealand's Kyoto Forest Credits? *New Zealand Journal of Forestry, 51*(2), 31.

Cole, C. V., et al. (1997). Global Estimates of Potential Mitigation of Greenhouse Gas Emissions by Agriculture. *Nutrient Cycling in Agroecosystems, 49*, 221–228.

Comerford, E. (2014). Understanding Why Landholders Choose to Participate or Withdraw from Conservation Programs: A Case Study from a Queensland Conservation Auction. *Journal of Environmental Management, 141*, 169–176. Available at: http://linkinghub.elsevier.com/retrieve/pii/S0301479714000784. Accessed May 12, 2014.

Commonwealth of Australia. (2012). *Deforestation and Reforestation Emissions Projections 2012*. Canberra.

Commonwealth of Australia. (2013). *Australia's Abatement Task and 2013 Emissions Projections*. Canberra.

Commonwealth of Australia. (2014). *Emissions Reduction Fund White Paper*. Canberra.

Council of the European Union. (2012). *Follow-Up to the 17th Session of the Conference of the Parties (COP 17) to the United Nations Framework Convention on Climate Change (UNFCCC) and the 7th session of the Meeting of the Parties to the Kyoto Protocol (CMP 7)*, 28 November. Durban, South Africa, Brussels: EN.

Cowell, S., & Williams, C. (2006). Conservation Through Buyer-Diversity: A Key Role for Not-For-Profit Land-Holding Organizations in Australia. *Ecological Management & Restoration, 7*(1), 5–20. Available at: http://dx.doi.org/10.1111/j.1442-8903.2006.00242.x.

Crossman, N. D., Bryan, B. A., & Summers, D. M. (2011). Carbon Payments and Low-Cost Conservation. *Conservation Biology, 25*(4), 835–845. Available at: http://dx.doi.org/10.1111/j.1523-1739.2011.01649.x.

CSIRO (Commonwealth Scientific and Industrial Research Organisation). (2003). *Assessing the Impact of Landcare Activities on Natural Resource Condition. Attachment 4* (Review of the National Landcare Program). Canberra.

Department of Sustainability, Environment, Water, Population and Communities. (2010). *Australia's Biodiversity Conservation Strategy 2010–2030*. Canberra.

Engel, S., Pagiola, S., & Wunder, S. (2008). Designing Payments for Environmental Services in Theory and Practice: An Overview of the Issues. *Ecological Economics, 65*(4), 663–674. Available at: http://www.sciencedirect.com/science/article/pii/S0921800908001420.

FAO UNDP UNEP. (2008). *UN Collaborative Programme on Reducing Emissions from Deforestation and Forest Degradation in Developing Countries (UN-REDD)*, UN. Available at: http://www.un-redd.org/AboutUNREDDProgramme/tabid/583/Default.aspx.

Garnaut, R. (2008). *The Garnaut Climate Change Review*. Melbourne: Cambridge University Press.

Gibbs, H. K., et al. (2007). Monitoring and Estimating Tropical Forest Carbon Stocks: Making REDD a Reality. *Environmental Research Letters, 2*(4), 1–13. Available at: http://stacks.iop.org/1748-9326/2/i=4/a=045023?key=crossref.4118e8af5a9a3ac02c1bb32f8a92c50f. Accessed March 5, 2013.

Hatton, T., et al. (2011). *State of the Environment 2011*. Canberra: Department of Sustainability, Environment, Water, Population and Communities. Available at: http://www.environment.gov.au/soe/2011/report/index.html.

Heller, N. E., & Zavaleta, E. S. (2009). Biodiversity Management in the Face of Climate Change: A Review of 22 Years of Recommendations. *Biological Conservation, 142*(1), 14–32. Available at: http://www.sciencedirect.com/science/article/pii/S000632070800387X.

Hill, T. (2011). UN Climate Change Conference in Durban: Outcomes and Future of the Kyoto Protocol. *Macquarie Journal of International and Comparative Environmental Law, 7*(2), 92–97.

Hulme, P. E. (2005). Adapting to Climate Change: Is There Scope for Ecological Management in the Face of a Global Threat? *Journal of Applied Ecology, 42*(5), 784–794. Available at: http://doi.wiley.com/10.1111/j.1365-2664.2005.01082.x. Accessed May 2, 2014.

IPCC. (2011). *Climate Change 2001: Impacts, Adaptation, and Vulnerability, Summary for Policymakers and Technical Summary of the Working Group II Report*. Geneva.

IPCC, & Cambridge University Press. (2007). Summary for Policymakers. In S. Solomon, D. Qin, M. Manning, Z. Chen, M. Marquis, K. B. Averyt, M. Tignor, & H. L. Miller, Eds., *Climate Change 2007: The Physical Science Basis. Contribution of Working Group I to the Fourth Assessment Report of the Intergovernmental Panel on Climate Change*. Cambridge: Cambridge University Press.

IPCC. (2014). Annex II: Glossary [Mach, K. J., S. Planton and C. von Stechow (Eds.)]. In *Climate Change 2014: Synthesis Report. Contribution of Working Groups I, II and III to the Fifth Assessment Report of the Intergovernmental Panel on Climate Change [Core Writing Team, R. K. Pachauri and L. A. Meyer (Eds.)]*. (pp. 117–130). Geneva, Switzerland: IPCC.

Kragt, M. E., et al. (2014). *What Are the Barriers to Adopting Carbon Farming Practices?* (Working Paper No. 1407). School of Agricultural and Resource Economics, University of Western Australia. Available at: http://ageconsearch.umn.edu/handle/195776.

Linacre, N., Kossoy, A., & Ambrosi, P. (2011). *State and Trends of Carbon Market 2011*. Washington, DC: World Bank.

Luck, G. W., et al. (2012). Ethical Considerations in On-Ground Applications of the Ecosystem Services Concept. *BioScience, 62*(12), 1020–1029. Available at:

http://bioscience.oxfordjournals.org/cgi/doi/10.1525/bio.2012.62.12.4. Accessed December 10, 2014.

Madsen, B., et al. (2011). *2011 Update: State of Biodiversity Markets, Offset and Compensation Programs Worldwide.* Ecosystem Marketplace. Available at: http://www.ecosystemmarketplace.com/pages/dynamic/resources.library. page.php?page_id=8393&section=our_publications&eod=1.

McCann, L. (2013). Transaction Costs and Environmental Policy Design. *Ecological Economics, 88,* 253–262. Available at: http://linkinghub.elsevier. com/retrieve/pii/S0921800912004958. Accessed November 13, 2014.

NASA Earth Observatory. (2012). *Seeing Forests for the Trees and the Carbon: Mapping the World's Forests in Three Dimensions.* Available at: http://earthob-servatory.nasa.gov/Features/ForestCarbon/.

Niesten, E., et al. (2002). Designing a Carbon Market That Protects Forests in Developing Countries. *Philosophical Transactions of the Royal Society of London, 360,* 1875–1888.

Pearce, D. (2005). Paradoxes in Biodiversity Conservation. *World Economics, 6*(3), 57–69.

Peters-Stanley, M., Katherine Hamilton, Marcello, T., & Sjardin, M. (2011). *Back to the Future: State of the Voluntary Carbon Markets 2011.* Ecosystem Marketplace & Bloomberg New Energy Finance.

Rode, J., Gómez-Baggethun, E., & Krause, T. (2015). Motivation Crowding by Economic Incentives in Conservation Policy: A Review of the Empirical Evidence. *Ecological Economics, 109,* 80–92. Available at: http://linkinghub.else-vier.com/retrieve/pii/S0921800914003280. Accessed November 19, 2014.

Sabine, C. L., et al. (2004). In C. B. Field & M. R. Raupach (Eds.), *The Global Carbon Cycle: Integrating Humans, Climate, and the Natural World.* Washington, DC: Island Press.

Sabto, M., & Porteous, J. (2011). Australia's Carbon Farming Initiative: A World First. *ECOS,* 160. Available at: http://www.ecosmagazine.com/ paper/EC10100.htm.

Spittlehouse, D. L., & Stewart, R. B. (2003). Adaptation to Climate Change in Forest Management. *FORREX-Forest Research Extension Partnership, 4*(1), 1–11.

Stephens, S. (2001). Visions and Viability: How Achievable Is Landscape Conservation in Australia? *Ecological Management and Restoration, 2*(3), 189–195. Available at: http://www.blackwell-synergy.com/links/doi/10.1046/ j.1442-8903.2001.00083.x.

UNFCCC. (2011). *Durban Climate Change Conference* (February 2012). Available at: First Commitment Period of Kyoto Protocol Will Phase Out in 2013 and Durban Climate Change Conference.

Vanclay, F., & Lawrence, G. (1995). *The Environmental Imperative: Eco-Social Concerns for Australian Agriculture.* Rockhampoton: Central Queensland University Press.

Van Noordwijk, M., et al. (2011). *How Trees and People Can Co-adapt to Climate Change Reducing Vulnerability in Multifunctional Landscapes.* Nairobi: World Agroforestry Centre (ICRAF).

Walker, S., et al. (2009). Why Bartering Biodiversity Fails. *Conservation Letters, 2*(4), 149–157. Available at: http://dx.doi.org/10.1111/j.1755-263X.2009.00061.x.

Wise, R. M., et al. (2014). Reconceptualising Adaptation to Climate Change as Part of Pathways of Change and Response. *Global Environmental Change, 28*, 325–336. Available at: http://linkinghub.elsevier.com/retrieve/pii/S095937801300232X. Accessed January 20, 2014.

CHAPTER 2

Adaptive Governance

Abstract This chapter presents a literature review about adaptive governance (AG) in natural resource management. It also explores a model for capturing challenges in achieving AG. The need for considering different stakeholders' voices and experiences is discussed in this chapter.

Keywords Adaptive governance · Knowledge exchange · Public learning · Programme design

ADAPTIVE GOVERNANCE DEFINITION

Top-down governance that works through regulatory processes to define environmental policies has limited ability to influence all actors and capture the complexity of different ecological systems and landscapes (Lockwood et al. 2010). This is because stakeholders have different power and urgency for their voice to be heard in the environmental decision-making process (Chaffin et al. 2014). In addition, from the ecological point of view, one plan does not fit all; some landscapes require urgent action and restoration. Adaptive governance (AG) has been introduced in response to the constant change of factors involved in environmental decision-making: climate change and land use change and to incorporate multiple actors' viewpoints (Dietz et al. 2003; Folke et al. 2005). *"Adaptive governance, involves the evolution of new governance institution capable of generating long term sustainable policy solutions to*

© The Author(s) 2019
N. Torabi, *Adaptive Governance in Carbon Farming Policies*,
https://doi.org/10.1007/978-3-319-97496-5_2

wicked problems through coordinated efforts involving previously independent systems of users, knowledge, authorities, and organised interests" (Scholz and Stifte 2005, p. 5). This means moving from the status quo and preparing institutions (both science and policy) and communities for dealing with the change and uncertainties.

The term AG was introduced by Gunderson (1999) suggesting the need for it in response to uncertainty and lack of balance in power among different stakeholders. Later, Dietz et al. (2003) continued the scholarly work focusing on the need for AG in dealing with the human nature interaction given the existing uncertainties. Since the introduction of the term, AG research has dramatically increased (Chaffin et al. 2014). Most of the theoretical and empirical work has focussed on resilience and socio-ecological systems. However, the demand for a paradigm shift in managing resources in both urban (Birkmann et al. 2010) and regional planning (Nelson et al. 2010) is evident. This shift considers new institutions and network of stakeholders to work together, moving away from the traditional top-down governance approaches.

ADAPTIVE GOVERNANCE IN AUSTRALIA

In an Australian context, Bryan et al. (2013) call for a "transformational adaptation" in Australian landscape to sustain within the environmental limits. This requires AG and new partnerships among multiple stakeholders. Dale et al. (2013) also emphasise the role of AG in the regional NRMs resource planning for a more sustainable landscape management in Australia. Nelson et al. (2010) state that the vulnerability of Australian rural communities to climate change needs to be addressed through AG. Leys and Vanclay (2011) argue that the social learning to build community capacity for AG is the key to managing competing objectives among multiple stakeholders in a forestry case study in Australia.

HOW TO MOVE FROM THE STATUS QUO?

To achieve AG, Scholz and Stifte (2005) discuss five challenges to build AG institution: representation, process design, scientific learning, public learning and problem responsiveness. Representation refers to the "who is involved" in the new institutions and procedures. Process design discusses "what mechanisms are in place" to ensure that decision-making considers all stakeholders and their needs. Scientific learning requires scientists and policymakers to work more closely together in natural resource management. This is to ensure that the introduced policy represents the scientific knowledge and learning behind it. Public learning involves engagement of the public in the decision-making process. It also indicates that communities need to familiarise themselves with the new process and respect the outcomes. Problem responsiveness is about the effectiveness of new policies in dealing with the existing natural resource management challenges while being fair and sustainable.

Adaptive governance requires different actors to work together and communicate effectively in the process of designing and implementing policies. This is especially essential when natural resource managers aim to engage landholders and land managers in sustainable practices to tackle ecological issues related to their landscape. Scientists, policymakers and community members require to actively discuss landscape-related challenges and solutions to achieve a consensus among themselves—a roadmap that would guide both policy and practice. However, it is worth noting that each group has their own priorities, outcomes and time frames. Landholders have their own priorities when managing their properties to run a business gain an income. Policymakers work under a tight time frame to deliver a policy by a certain deadline. Scientists also have different tools to measure the success of their project delivery (e.g. scientific papers and reports). In addition, each group communicates in a different way and method. These differences could result in a mismatch between socio-ecological challenges and the solutions provided by policymakers or scientists. These solutions are then translated to polices or land management schemes. Many of these schemes require landholders to make permanent shifts to their properties, for example, revegetation. For landholders to undergo the changes, they need to be part of the decision-making process.

Knowledge exchange between science and policy is essential in achieving AG in the conservation realm (Cook et al. 2013). Adopting more innovative approaches for scientists, policymakers and funding bodies such as providing resources and also recognising the importance of it could assist the success of knowledge exchange (Cvitanovic et al. 2015). Lockwood et al. (2010) refer to knowledge exchange in their nine principles (refer to principle 7: capability) of AG in the NRM context. This is because knowledge is a key component of dealing with uncertainty and change in ecosystems (Lockwood et al. 2010). The role of boundary or bridging organisation in filling the gap in the knowledge exchange and improvement in governance has been highlighted (Moloney et al. 2018). These organisations connect different actors and promote knowledge utilisation in AG (Crona and Parker 2012).

The need to hear different stakeholders' voices and insight is vital to achieving AG. This is because multiple knowledge sources, efforts and institutions working together could help achieve a more sustainable outcome. In doing so, I use Scholz and Stifte's (2005) "five challenges model" to examine how close biodiverse carbon planting is from AG. I investigated the science, policy and public stakeholders' opinions and experiences about each challenge in practice through face-to-face interviews. This provides decision-makers to have an insight into the differences between theory and practice of AG in carbon farming. By representing this gap, policymakers could take initial steps towards institutions that facilitate AG in carbon farming and indeed in different aspects of natural resource management. Chapter 4 explores the gap between theory and practice in AG by focusing on each element of the model based on the multiple stakeholders' opinions.

REFERENCES

Birkmann, J., et al. (2010). Adaptive Urban Governance: New Challenges for the Second Generation of Urban Adaptation Strategies to Climate Change. *Sustainability Science, 5*, 185–206.

Bryan, B. A., et al. (2013). The Second Industrial Transformation of Australian Landscapes. *Current Opinion in Environmental Sustainability, 5*(3–4), 278–287.

Chaffin, B. C., Gosnell, H., & Cosens, B. A. (2014). A Decade of Adaptive Governance Scholarship: Synthesis and Future Directions. *Ecology and Society, 19*(3), 56.

Cook, C. N., et al. (2013). Achieving Conservation Science That Bridges the Knowledge-Action Boundary. *Conservation Biology, 27*(4), 669–678.

Crona, B. I., & Parker, J. N. (2012). Learning in Support of Governance: How Bridging Organizations Contribute to Adaptive Resource Governance. *Ecology and Society, 17*(1), 32.

Cvitanovic, C., et al. (2015). Improving Knowledge Exchange Among Scientists and Decision-Makers to Facilitate the Adaptive Governance of Marine Resources: A Review of Knowledge and Research Needs. *Ocean and Coastal Management, 112*, 25–35. Available at: http://dx.doi.org/10.1016/j.ocecoaman.2015.05.002.

Dale, A., et al. (2013). Carbon, Biodiversity and Regional Natural Resource Planning: Towards High Impact Next Generation Plans. *Australian Planner, 50*(4), 328–339.

Dietz, T., Ostrom, E., & Stern, P. C. (2003). Struggle to Govern the Commons. *Science, 302*(5652), 1907–1912.

Folke, C., et al. (2005). Adaptive Governance of Social-Ecological Systems. *Annual Review of Environment and Resources* (pp. 441–473). Palo Alto: Annual Reviews.

Gunderson, L. (1999). Resilience, Flexibility and Adaptive Management— Antidotes for Spurious Certitude? *Conservation Ecology, 3*(1), 7.

Leys, A. J., & Vanclay, J. K. (2011). Social Learning: A Knowledge and Capacity Building Approach for Adaptive Co-management of Contested Landscapes. *Land Use Policy, 28*(3), 574–584. Available at: http://linkinghub.elsevier.com/retrieve/pii/S0264837710001171. Accessed July 15, 2014.

Lockwood, M., et al. (2010). Governance Principles for Natural Resource Management. *Society and Natural Resources, 23*(10), 986–1001.

Moloney, S., Bosomworth, K., & Coffey, B. (2018). "Transitions in the Making": The Role of Regional Boundary Organisations in Mobilising Sustainability Transitions Under a Changing Climate. In T. Moore et al. (Eds.), *Urban Sustainability Transitions. Theory and Practice of Urban Sustainability Transitions* (pp. 91–108). Singapore: Springer. Available at: https://doi.org/10.1007/978-981-10-4792-3_6.

Nelson, R., et al. (2010). The Vulnerability of Australian Rural Communities to Climate Variability and Change: Part II-Integrating Impacts with Adaptive Capacity. *Environmental Science and Policy, 13*(1), 18–27.

Scholz, J. T., & Stifte, B. (2005). *Adaptive Governance and Water Conflict: New Institutions for Collaborative Planning.* Washington, DC: Resources for the Future.

CHAPTER 3

Understanding Stakeholders:
Awareness of Carbon Farming Schemes

Abstract Torabi explores the factors that impact awareness about biodiverse carbon schemes, presenting the findings from the socio-demographic survey and landholders' interviews. This chapter explores the first two steps of adoption theory and explores landholders' motivations and barriers in each stage. These findings could assist policymakers in progressing towards adaptive governance in the carbon farming policies.

Keywords Adoption theory · Socio-cultural drivers · Motivations Biodiverse carbon plantings · Awareness step

INTRODUCTION

This chapter explores the role of landholders' sociocultural drivers in awareness of carbon farming schemes. Both landscapes *and* landholders in rural areas are affected by climate change. The former is well recognised within the disciplines of ecology and environmental science, and the ecological benefits of revegetation for both carbon sequestration and biodiversity conservation are well studied (Hulvey et al. 2013; Standish and Hulvey 2014). Some impacts on landholders are also well studied, for example financial (Lin 2011; Rochecouste et al. 2015), social and health-related impacts (Addison 2013; Adger et al. 2005). However, one impact on landholders that is relatively understudied is the growing need for them to participate in mitigation projects, such as private land conservation schemes. This chapter aims to understand how landholders have responded to the growing pressure to participate in schemes, gain insight into the social and cultural factors driving their involvement in biodiverse carbon planting and predict the uptake of schemes based on pre-existing drivers. *"You cannot save the land apart from the people or the people apart from the land. To save either, you must save both"* (Wendell Berry 1995, p. 56). I take the position that techno-political issues, such as setting targets for carbon abatement in Australia, are inextricably intertwined with socio-ecological systems, for example how primary land use objectives are balanced and managed with new conservation schemes in an inhabited landscape. Yet policies are typically set with solely ecological targets, for example, to reach the GHG abatement of 131 $MtCO_2$-e (Commonwealth of Australia 2013). An unintended consequence of this is that the human agents whose efforts and actions will determine the policies' success or failure often go unacknowledged. Encouraging landholders to adopt new land management practices is an important element of the design of biodiverse carbon sequestration schemes.

DATA COLLECTION

Data were collected using an initial survey and then through interviews with private landholders who are participating in a biodiverse carbon planting scheme in Victoria, with the aim of exploring the social and cultural drivers of participation in bio-sequestration projects. I chose to only survey and interview landholders who were already participating in the planting scheme. This is because the primary aim was to explore what results in participation in each step of programme adoption. While including non-participants could have helped to identify obstacles to the early stages of adoption (e.g. awareness and interest), by definition their non-participation in the scheme meant that they had no experience in the later stages of adoption (e.g. adoption and post-adoption). Including non-participants would therefore have presented an uneven focus on the early stages of adoption. In addition, accessing non-participants presented logistical problems. The non-participant population was much harder to access—my initial attempts to make contact identified only two landholders in this category. In total, I surveyed and interviewed 17 landholders and interviewed 14 other stakeholders (scientists and policymakers).

Survey, Interviews and Observational Studies with Landholders

Initially, surveys were distributed as they provide a broad understanding of the research sample. A range of open and closed questions (including some scales, yes–no) was included to obtain a better understanding of landholders' demographic and socio-economic profiles and their environmental concerns. Bryman (2004) argued that closed questions (e.g. Have you ever participated in any conservation activities?) have a role to play in collecting factual and demographic information. Closed questions are more convenient for the researcher to process, whereas with open questions rich responses can be expected (Bryman 2004; Dohrenwend 1965). This is because closed questions offer "fixed choices" for the participants (Balnaves and Caputi 2001, p. 78). The survey was mailed by Greenfleet to all private landholders who participated in biodiverse carbon plantings on their properties in Victoria. A series of demographic questions sought to obtain information about landholders' ages, education and property size. In addition, the survey included questions about the value landholders placed on co-benefits of biodiverse carbon plantings (e.g. the important factors influencing planting those trees and the value of trees on their properties).

My rationale for conducting in-depth semi-structured interviews was to obtain comprehensive individual data about their experiences, perceptions and opinions on which to build a more credible social conceptual model. Bryman and Burgess (1999) state that interviews can be considered as "special conversations" about people's experiences (Holstein and Gubrium 2003). In-depth semi-structured interviews could lead to the understanding of "social actors' meanings and interpretation" of their involvement with the studied phenomena (Blaikie 2000, p. 234). Semi-structured questions allowed the clarity of process to be an unfolding evolution through a team of participant and researcher.

To obtain an in-depth understanding of the context (in this case, the sociocultural drivers of the landholders), face-to-face, semi-structured interviews were undertaken. Survey participants interested in a one-to-one interview were requested to contact me. Individual interviews (17 applicants) were held at the participants' properties between January and September 2013. Each interview took from 90 to 120 minutes. Interviews continued until I reached the data saturation point (Glaser and Strauss 1967) where no new themes were emerging. Interviews were tape-recorded and transcribed verbatim.

I prepared an "interview guide" (Bryman 2004) consisting of questions about the process by which the participants joined Greenfleet and their motivation for doing so, their knowledge of the carbon and biodiversity markets, their ideas about opportunities to integrate both markets, the degree to which biodiversity is valued, and their likely responses to integrated biodiversity and carbon planting policies. To capture their stories, I started asking about the history of their property and conservation activities, their experiences with nature, and conservation activities as children. Later, I asked about their current land management practices, the process of joining the voluntary biodiverse carbon planting and their future planned conservation activities.

A few examples of interview questions are listed below.

- Let's talk about your property.
- How did you come across Greenfleet? [Did someone recommend them to you?]
- Have you recommended plantings to anyone else? If yes, have they commenced biodiverse carbon plantings?
- Have you been involved in any other conservation schemes (land care, Bush broker)?
- Were there any changes associated with management of your property since these trees were planted?
- What made you so passionate about the environment?

After undertaking the in-depth semi-structured interview, I walked through landholders' properties and carbon planting sites to help me better understand the study area. My rationale was to have an opportunity for closer observation of my case study sites (Blaikie 2000). Together with the landholders, I walked through their properties and gathered field notes, spending time in their "naturalistic" setting (Cooper et al. 2009). This helped me to gain a better understanding of their sense of biodiversity and their feelings towards the carbon planting itself that they could not explain during the interview. This also assisted me to find out why and how they chose the spatial position of the carbon plantings.

This chapter begins with a description of participating landholders' profiles and their socio-demographic characteristics. It provides a portrait of interviewees whose participation will be illustrated in detail. After introducing the research participants, I apply adoption theory (Pannell et al. 2006) to examine the drivers for the first two steps of participation: awareness and non-trial evaluation.

PROFILE AND DEMOGRAPHIC CHARACTERISTICS OF LANDHOLDERS

As this chapter focuses on the landholders' voice and lived experiences, a useful starting point is to characterise their profiles and sociodemographics. Table 3.1 contains landholders' pseudonyms adopted for this study, the Catchment Management Authority (CMA) in Victoria within which their properties are located, the type of land use they associate themselves with and the size of their biodiverse carbon plantings.

I have identified properties within CMA boundaries as conservation plans and policies are designed and implemented within those boundaries. In the "awareness" section of this chapter, I will reflect on the various large landscape and biodiversity conservation plans in each CMA and the link between participation and awareness of such plans. Figure 3.1 presents the location of CMAs in which surveys and interviews were conducted.

Table 3.1 Landholder profiles: Pseudonym, type of land use, their CMA and plantation size

Name	Land use	CMA	Plantation size (ha)
Steve	Cattle grazing	North Central	40
Laura	Lifestyle	North Central	37
Noah and Linda	Lifestyle (dog breeding)	North Central	28
John	Cattle grazing	North Central	8
David	Cattle, sheep grazing	North Central	6
Mat	Wool, beef, lamb	Goulbourn Broken	9
Andrew	Cattle grazing	North Central	2.5
Luke	Cropping, sheep and cattle grazing	North Central	47
William	Lifestyle	West Gippsland	15
Barbara	Lifestyle	North Central	19
Anna	Lifestyle	Goulbourn Broken	7
Ryan and Owen	Wool, cattle grazing	North Central	2.7
James	Cattle grazing	West Gippsland	8
Jacob	Recreation-tourism	West Gippsland	43
Oliver	Wool, beef, lamb	Glenelg Hopkins	189
George	Cattle grazing	West Gippsland	10
Daisy	Lifestyle	North Central	25

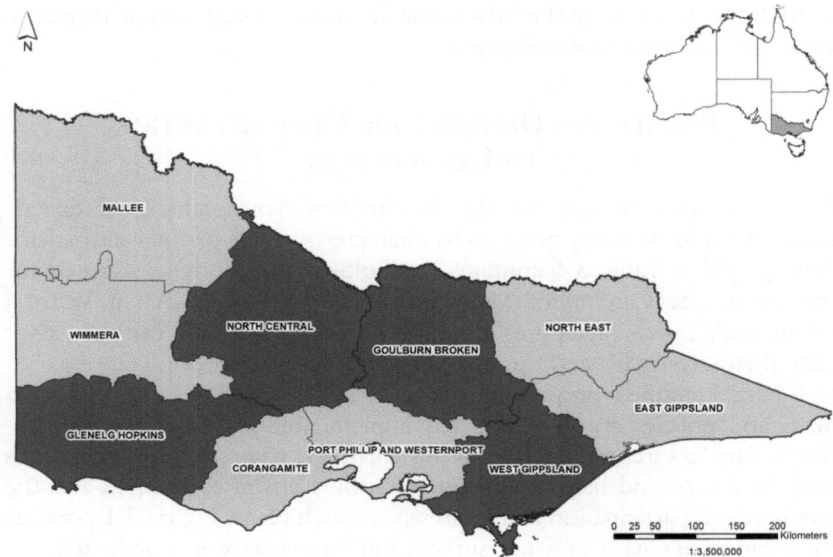

Fig. 3.1   CMAs in which the research is conducted

Table 3.2 Socio-demographic characteristics of landholders

Age groups (year)	Number of respondents	%
18–24	0	0
25–39	3	18
40–54	7	41
55–69	5	29
More than 70	2	12
Land use		
Commercial	7	41
Semi-commercial	3	18
Hobby farmer	2	12
Lifestyle landholder	5	29
Source of income		
Off-farm	10	59
On-farm	7	41
Level of education		
High school	3	18
TAFE	2	12
University	12	70

Socio-demographic features of landholders who participated in this research were obtained through a written survey, as described in Chapter Two. Seventeen people participated in the initial survey. They were interviewed in detail in the follow-up stage of the research. Age group, type of land use, type of income and education level are presented in Table 3.2. Most participants were 40–54 years old (41%) and run commercial and semi-commercial properties (59%). In this study, 59% of the respondent landholders rely on off-farm income sources. Furthermore, 70% of them have a university education.

FROM AWARENESS TO POST-ADOPTION OF BIODIVERSE CARBON PLANTINGS

My objective in this section is to explore the factors influencing landholder participation in biodiverse carbon plantings at first two steps of the adoption process. Empirical data will be presented, along with the supporting literature as the chapter progresses.

In natural resource management, the adoption process has been shown to be unpredictable, uncertain and nonlinear (Morris et al. 2000; Öhlmér 1998). Political uncertainty and institutional changes can have negative impacts on the likelihood of adoption (Morris et al. 2000), and low participation rates will reduce the anticipated ecological benefits (Comerford 2014). This leads to inefficient spending (time and money) by the responsible institutions and may create negative public perceptions of other policies to be introduced in the future (Mankad et al. 2015). Hence, it is essential to understand the adoption process and landholders' social and cultural values at each step, and how these values affect their decisions to participate in biodiverse carbon schemes.

STEP ONE: WHAT ARE THE IMPORTANT ELEMENTS IN RAISING AWARENESS AMONG LANDHOLDERS?

The awareness phase refers not only to the knowledge and recognition of the existence of a programme, but to the degree to which landholders perceive it as feasible and applicable to their current business (Pannell et al. 2006). In addition to landholder characteristics, context (e.g. social settings and norms in the community) and external factors (e.g. policy settings) were specified in the literature as influential factors in the awareness phase (Morris et al. 2000).

Cultural Motivation

Culture is worth a second look, because it envelops both problems and solutions (Rankin 2014, p. 15).

Cultural motivation is one of the prior conditions described by Morris et al. (2000) that facilitate the early stages of adoption. Cultural capital is formed through social norms, laws and ideologies (Pannell et al. 2006) together with other human factors that influence behaviour, such as skills and experiences (Burton and Paragahawewa 2011). Cultural legacy is another potential motivator and can be translated through considering land as a family asset (Fischer and Bliss 2008). As a direct question about culture seems intangible to many, I used the family tradition of conservation and childhood experiences as a proxy to define cultural elements. Almost 60% of landholders surveyed in this study grew up on a farm and undertook conservation activities like tree planting in their childhood. This cultural motivation possibly helped some landholders to take their first steps towards biodiverse carbon plantings.

Two important elements of cultural experience were observed in this research. First, having a role model and social learning from the very early stages of life was a key element driving participation. Family influences have been recognised as a significant factor in conservation practices (Chawla 1999). John recognises his father's conservation work: *"Dad did a lot of soil conservation work. Did a lot of tree planting too…"*. Laura also expressed her father's beliefs about revegetation and balance in the agricultural landscape as her role model.

> … Dad was before his era. He already believed that there had to be a balance, like nature helped us. He always encouraged birds and certain wildlife as he felt that they were a benefit in some ways to have around as well as just being good to look at. So he would have agreed with some revegetation. (**Laura**)

Second, the opportunity of growing up on a farm and building a connection with nature is a major driver for some participants. Daisy reflected on her experience as a child growing up in the bush and the willingness to restore and conserve native species.

> We both grew up in the bush and we just know how important it is to bring it back, so to have that opportunity to plant something that will suit the area so it will grow better because it's always grown here, and then

it also has those benefits of yeah, providing a habitat. And we wanted to enjoy the habitat you know knowing we wanted to have kids, we wanted to be able to see the kangaroos and the birds and the honey-eaters

This echoes the relevant literature about the importance of nature encounters and childhood learning about environmental conservation (Chawla 1999; Louv 2010). Anna also expressed her passion for conservation activities as a result of her childhood experiences in nature and their rural property: "*So we would go up there [family's rural property] every weekend and look after the cows and do fencing and put out hay and all those things that you do... Ride ponies, so I was very much connected to landscape*" (**Anna**).

The interview responses illustrate that landholders gathered knowledge and values as they grew up and built up a culture for conservation within them. "*I also have just knowledge from growing up the way that I did*" (**Anna**). It echoes the array of literature representing the role of nature in children's well-being and learning process (Kahn and Kellert 2002; Kahn et al. 2009; Louv 2010). Such exposure to nature and conservation activities encouraged them to participate more in environmentally beneficial land management practices as adults.

Formal and Informal Education

Many studies identify the impact of formal education on participation in conservation activities (Comerford 2014; Ecker et al. 2012; Ma et al. 2012). The probability of participation in conservation activities is higher for landholders with higher levels of education (Ahnström et al. 2008). As indicated in Table 3.2, 82% of participants have TAFE qualifications or university degrees. Landholders' reflections on the role of formal education in raising their awareness about private land conservation practices are documented in this study. "*Well, the Whole Farm Planning Course gave me a really good base*" (**William**). Research also reveals that it is not only landholders' education that is significant but also their spouses' level of education that is related to their participation in community-based natural resource management programmes like Landcare (Ecker et al. 2012). As Barbara mentioned, "*Well, my husband's actually doing a Bachelor of Agricultural Science at the moment. So he gets quite a lot of information that we use from his course as well as just using the web*" (**Barbara**). This emphasises the importance of formal education in the conservation-related field for raising awareness.

Kilpatrick and Johns (2003) argue that in addition to formal education, informal learning (e.g. other farmers or friends) is an essential factor in increasing landholders' knowledge of sustainable land management. Inspirational figures, community members, neighbours, reading various related sources and Landcare groups are among the various informal learning sources mentioned by landholders in this study. Mat, one of the landholders, mentioned the presence of an inspirational figure regarding land conservation practices in the early stages of his life.

> I used to jackaroo. This was just back in the seventies, and my boss there, he'd just been made the Landcarer of the month, but he's got a place you would have come past, and he's .. I suppose he had an influence on me when I was younger, and you know the place that he manages now, he's very much the son, he's an inspiration.. and he's fifteen years older than me, but he's been doing it and in a lot harsher environment than me, but you know, he's one who's done a lot, yeah. (**Mat**)

Other landholders mentioned different methods of receiving information and becoming aware of new schemes in their area. Anna mentioned some of these conduits, such as courses, book and related government websites and local farmers.

> I did some day courses through Greening Australia about seed collection and propagation. I attend quite a number of things, and I talk to quite a lot of people and so I garner information from all of those sources. I also have a collection of books that I get material from, and I also look from time to time, especially around things like weed management, I'll look on the DPI website or talk to the local farmers. (**Anna**)

This finding echoes previous research about informal learning; Latchem (2014) argues that 70–90% of human beings' learning is informal.

Social learning in the workplace and being engaged with conservation practices in natural resource management is influential in understanding social-ecological systems (Pahl-Wostl et al. 2008). Professional experience as part of informal learning was observed in some interviews. Some participants stated that their knowledge partly came from their involvement in similar activities in their profession. "*I was working for the [...] Catchment Management Authority, so I had the role of bush care facilitator and had the task of bringing out a draft native vegetation plan for the region*" (**George**).

Emergent Stewardship: Conservation Activities on the Property

Land stewardship values may be developed in landholders once they care for their property beyond consideration of farm benefits and monetary values. Such landholders are more likely to participate in tender-based conservation schemes (Blackmore and Doole 2013). Landholders also can develop a sense of stewardship as a result of their everyday agricultural practices (Trigger et al. 2010). In addition, undertaking conservation activities independently (i.e. revegetation) makes biodiverse carbon plantings fit with existing land management or farming practices (GFIT) (Blackmore and Doole 2013; Pannell et al. 2006; Wilson and Hart 2001).

Many of interviewed landholders were undertaking different conservation activities and practices on their properties before engaging with biodiverse carbon planting schemes. Based on the initial survey, these activities are listed in Table 3.3. As this table illustrates, most participants were actively engaged in revegetation practices.

The interview materials also demonstrate some of the conservation practices undertaken by landholders. Mat reflects on the type of work he has been doing to run his agricultural business. *"We've done a lot of native pasture regeneration, so that's sort of managing how we graze the paddocks to encourage native pastures as opposed to exotic introduced ones, and we've done a lot of recovery work, there were a lot of blackberries when we first came here..."* (**Mat**). This keeps landholders constantly engaged with conservation activities and seeking information regarding any new practices and schemes (Cooke and Lane 2015). Interestingly, John calls the soil enhancement practices part of agriculture rather than real conservation activity.

Table 3.3 Conservation activities undertaken by landholders

Conservation activity	None	Some	Quite a bit	An extreme amount	Total responses
Controlling weeds	0	3	8	5	16
Controlling pest animals	1	2	10	2	15
Fencing re-vegetated areas	1	1	8	5	15
Erosion control	0	6	7	2	15
Re-vegetation	1	0	9	6	16
Grazing strategies	1	7	5	3	16
Others: Ecological burns, farm forestry	5	0	0	1	6

I've done a lot of soil erosion work... yeah, and we've done a lot of other plantings that isn't carbon-related. Yeah. And there's still more to be done, but yeah... Soil enhancement, but I mean in terms of putting lime on to decrease acidity and stuff like that, but you know that's really agricultural rather than conservation. (**John**)

Andrew also undertakes conservation activities on his property as a means to improve environmental conditions.

The creeks were all washing out and I planted all along the creeks thousands of trees. There's 1800 meters along the back – we did the same because there's a creek next door that's cutting into my place, and so we planted all the trees along there and it stops erosion as well as creates birdlife and corridors.

Emergent stewardship was observed in the course of this research. It reveals that landholders start conservation activities on their properties as a means to survive and prosper in their agricultural business. Once they become proficient in such activities, they search for new practices and innovative ways to conserve the landscape. This provides opportunities for experimental learning that could develop conservation values (e.g. emergent stewardship) among landholders (Cooke and Lane 2015). It echoes the research undertaken by Trigger et al. (2010) that focuses on the emergence of stewardship values as a result of undertaking conservation practices embedded into agricultural business.

In addition to undertaking land management practices on their properties, landholders have participated in conservation programmes like Bush Tender and Trust for Nature. Stoneham et al. (2003) state that such programmes alter environmental awareness among communities. This constant engagement with ecology also reflects the emergent stewardship concept (Cooke and Lane 2015).

Related Conservation Schemes in the Area

Some landholders were aware of the presence of carbon and/or biodiversity conservation activities in their CMAs. They undertook plantations as they saw themselves as a part of a larger landscape connectivity programme in the area. As John expressed it: "*There's some connection here. We did have a big project which was to link.... it was originally conceived as part of a much bigger project which was the Campaspe [river] to the Cobaw [State forest] project"*. This reveals that landholders were feeling connected to a bigger picture in the landscape. This awareness and participation in related conservation practices have provided landholders with the opportunity to develop stewardship values, as supported by previous research (Ahnström et al. 2008).

Furthermore, awareness and raising interest could happen when CMAs are active in promoting their carbon- and biodiversity-related schemes among landholders (Meadows et al. 2014). This can assist landholders to keep up with information about other new land conservation schemes (e.g. biodiverse carbon planting).

Table 3.4 shows the relevant conservation activities in the study area.

Participation in a Landcare Group

Landcare groups are perceived as "community process based on a learning group" (Martin 1997, p. 51). They play a role in empowering social capital, assisting landholders to learn about new land management practices and acting as a conduit for communication (Compton and Beeton 2012; Sobels et al. 2001). Fifty-three per cent

Table 3.4 Related conservation activities in the CMAs within which interviews took place

CMA	Conservation (carbon and/or biodiversity) plans in the CMA	Description
North Central	Kyneton woodlands	To provide incentives for landholders to undertake biodiverse carbon stores on their properties
West Gippsland	Red gum grassy plains	To undertake the recovery of Red Gum Grassy Plains through Caring for Our Country (CFOC) funding
Goulburn Broken	The Broken Boosey Conservation Management Network (CMN)	To protect Box Ironbark forests
	Climate change strategy	Support mitigation and sequestration activities aiding from Biodiversity Fund, Land sector package
Glenelg Hopkins	Connecting the catchments	To manage environmental threats (e.g. by vegetation along creeks)

of landholders were involved in their local Landcare group. They attend it to receive information regarding grants and land management initiatives and to get engaged in their community socially. As Andrew noted:

> Well, I belong to the […] Landcare and in fact, I was in it for about ten years, eleven years, backed out now, but I'm nearly 75, and the governments were still giving grants out so we all applied for different grants. And fortunately we got a fairly good grant which did a lot of the fencing; I couldn't have afforded to do it otherwise.

Some landholders also received information about biodiverse carbon plantings from Landcare presentations. As James mentioned: "*When I was president of Landcare, they wanted to come and talk to the group. Well, most of the neighbours and things were in Landcare so they all were at the presentation by Greenfleet*" (**James**). This finding is aligned with previous research about the higher rate among landholders who belong to a Landcare group choosing to participate in private land conservation activities (Jellinek et al. 2013). However, Compton and Beeton (2012) argue that Landcare leaders' skills and experiences as a change agent enables landholders to move from the status quo and participate in conservation programmes.

Social Networks and Trusted Peers

Collaborative interactions among landholders to share knowledge and resources have been well documented (Lauber et al. 2008). Social connectedness assists landholders in rural communities to access information regarding conservation practices (Baumgart-Getz et al. 2012). They can keep their knowledge about conservation activities up to date, which may lead to higher rates of programme uptake (Morrison et al. 2008). Connectedness of landholders to early adopters of carbon farming practices increases participation rates among landholders (Kragt et al. 2014). The role of social capital (trust and social networks in particular) at different stages of biodiverse carbon planting adoption (and post-adoption) is unpacked further in Torabi et al. (2016).

Second Step: Non-trial Evaluation of Biodiverse Carbon Planting Schemes by Landholders

Once landholders become aware of the scheme through their social networks or through seeking advice about other conservation practices on their properties, they enter the non-trial evaluation stage (Pannell et al. 2006) of biodiverse carbon planting. In this phase, landholders appraise their drivers for participation. They also consider the compatibility of the new practice to their property management (Blackmore and Doole 2013). Increasing farm productivity and the environmental co-benefits are among the drivers that accelerate landholders' participation in private land conservation schemes (Jellinek et al. 2013).

The primary aim of policies like the carbon farming initiative (CFI) or any other monetary or paid carbon farming scheme is to sequester carbon from the atmosphere (Australian Government Clean Energy Regulator 2014). Carbon sequestration and emission abatement are the principal messages communicated within policy documents and with landholders. In the course of this research, carbon sequestration was not mentioned by any landholder as a primary driver for participation in biodiverse carbon planting schemes. Landscape-related co-benefits were unambiguously the major drivers for landholders to participate in these programmes.

Co-benefits of Biodiverse Carbon Plantings

"So there's a range of reasons. So there's the economic and production side; there's the biodiversity side, and then on top of that there's also the land repair side I suppose, for want of a better world" (**Oliver**). As summarised in this quote, there is a range of reasons for participating in biodiverse carbon plantings: productivity gains, biodiversity conservation and land rehabilitation are among the main drivers why landholders participate in such schemes (Jellinek et al. 2013).

Biodiversity Conservation
Biodiversity Conservation Working Alongside Agriculture
Landholder interviews reveal the importance of biodiversity conservation as one of the drivers for landholders to undertake biodiverse carbon plantings on their properties. Participants recognised preserving the diversity

of species as one of the co-benefits that such re-vegetations would have. *"Well probably the main benefits are biodiversity benefits. So we've got a sort of a commitment and a desire to see conservation and agriculture both working"* (**John**). John, one of the landholders, explains that balancing agriculture productivity or amenity benefits with conservation seems critical to landholders. It also reveals that they do not consider conserving biodiversity as a barrier to agricultural productivity, recognising agriculture as an "ecological enterprise" (Saunders and Walker 1998).

Market-based instruments (MBIs) have been extensively used in agro-environmental contexts (Moon and Cocklin 2011; Morrison et al. 2008; Doole et al. 2014). They provide financial incentives to landholders to undertake desired activities (Stern 2008). Jack et al. (2008) argue that, in designing a successful incentivising programme, socio-environmental contexts need to be considered. In doing so, this study reveals that landholders are not necessarily profit maximisers and they have a strong commitment to conserve the landscape. *"But right at the start we were going to protect the bush anyway, whether it cost us or not"* (**George**). This illustrates that stewardship motivations can be stronger than monetary ones. This echoes Kragt et al. (2014) who found that the financial incentives are not the main drivers for the uptake of carbon plantings among landholders in Western Australia.

Recognising Intrinsic Values: Mixed Species Value on Its Own and as Habitat for Wildlife

The value of mixed species revegetation to halt flora extinction and as a habitat for wildlife has been the subject of previous research (Bigsby 2009; Bowen et al. 2007; Hartley 2002). However, in addition to recognising and appreciating biodiversity as mixed species (Saunders and Walker 1998), the habitat they provide for wildlife is valued intrinsically (Lockwood 1999) by landholders. *"Well you know the whole country needs it in the sense that if you don't have the mixed species, what does the birdlife and the animal life live on? You're limited in what .. in the corridors of what animals can exist"* (**Andrew**). As Andrew explained, the biodiverse carbon plantation provides several ecological benefits to fauna and flora on his property that he values intrinsically. In a similar example, Luke emphasises that increasing wildlife abundance has been one of the drivers to participate in such schemes. *"We were also interested in creating an increase in wildlife through the farm and providing a habitat for that, which is always good to see a bit of birdlife and that sort of thing"* (**Luke**).

Habitat Corridors and Landscape Connectivity

Intrinsic values for the ecosystem as a whole and considering broader landscape restoration are considered "ecocentric values" (Lockwood 1999). These types of values and ethical considerations could refer to the sense of stewardship when landholders think about public goods beyond their own properties' boundaries. When I asked Laura about her motivations in the non-trial evaluation phase, her response revealed both the intrinsic values for biodiversity and habitat restoration, and aggregating her local action to a broader restoration context in the landscape (Menz et al. 2013).

> One end of my property is a small Crown area of bush that was maybe an old school reserve or something, and then on the other side we have another one, and in the middle of my dam, dad has always maintained these tree groups anyway for birdlife and so forth, so I thought if I re-vegetate and try and join the whole lot up so that will then give birdlife and perhaps the smaller animals a chance to get from one place to another. (**Laura**)

The ability of landholders to see themselves in a broader picture of the landscape and the desire to be part of larger landscape connectivity programmes could assist in the success of the latter. Such programmes often require actions on a broader scale across the landscape (Briggs 2001).

Land rehabilitation is another driver for landholders to consider participation in biodiverse carbon plantings. The aspiration to restore landscapes reflects the ethical consideration and stewardship values that landholders hold. As Daisy explained, some of the landholders had the idea of restoring the land to provide a more suitable habitat for the native fauna. *"We just.. we wanted to rehabilitate the land I suppose, so we wanted to bring back what would have been here with the habitat to the local fauna"* (**Daisy**).

Another key point that can be gleaned from landholders' comments is the fact that they tend to hold a "living with nature" approach towards land (Thompson et al. 1994). *"So it's just we kind of feel like that that we're doing something for the health of the land and giving something back instead of just taking stuff away from it"* (**Barbara**). Other landholders like Barbara reflected an ecocentric approach (Buijs 2009) to their property management. This approach encompasses moral aspects of caring for land and re-emphasises the findings of Gill (2013) who considers stewardship as an ethical element of land management.

Farm-Related Drivers

These groups of drivers refer to factors that are mainly related to farm productivity. Increasing farm productivity has been documented as one of the drivers for landholders to participate in private land conservation (Ahnström et al. 2008; Jellinek et al. 2013). Windbreaks and shelterbelts to assist the livestock in extreme weather conditions, erosion and salinity control are among these motivations.

Windbreaks and Shelterbelts

One of the drivers landholders mentioned was to create windbreaks and shelter belts for the benefit of their stock. This is not only a driver but a determining factor for choosing where to spatially plant the trees. In the course of the interviews (particularly while walking/driving around their properties), I asked landholders about the spatial location of trees they planted and the justification for the choice of location, ecological and biophysical explanations. In addition to strategically locating them in a way to connect to the Crown land or existing adjacent remnant vegetation (discussed earlier in this chapter), the prevailing direction of wind and protection of livestock was an important factor for landholders. "We've put them there more or less for wind, because we get all the wind from the south-west, and in winter it's cold so you create shelter belts for the stock" (Andrew). David also recognises the fact that windbreaks are essential in the sheep grazing areas.

> It's definitely a plus for sheep people that the biggest loss of mortality of lambs is not having shelter... like it's cold wind and rain together is the biggest lamb killer. So if you have shelter from that exposure, and whether that event comes and you're lambing, if you happen to be lambing right then and it's windy and rainy, you'll lose a lot of lambs, and the shelter would prevent a lot of that. (**David**)

Erosion and Salinity Control

Erosion and salinity control have implications for on-farm conservation and also benefits for a broader landscape health (Johnson et al. 2007). From a landscape restoration point of view, Steve reflected that biodiverse carbon plantations on his property play an important role in absorbing salt from the downstream lake. *"It's right at the top of the catchment for Lake [...], and we're saving up a lot of the water so fighting the salt and we have the benefit of grazing"* (**Steve**). From an on-farm perspective, many landholders mentioned similar points, including Oliver: *"I suppose there's also where we've had... There's some salinisation issues we've had at different areas which we've actually redressed by planting trees"*. Fighting erosion, as Noah and Linda mentioned, has been an important driver for landholders to re-vegetate their properties by up taking a biodiverse carbon planting scheme. *"We would like to think that the planting we did would help with the erosion in this area, and I think that has helped because there was a fair bit of erosion – it's a fairly dry area, as you probably gathered. There's a fair bit of erosion around"*. Again, tackling erosion is not only considered an on-farm related issue, but it is important from the perspective of restoration of a broader landscape. William explained his passion for fighting erosion and restoring landscapes as follows. *"The erosion is shocking and I mean the magnificent trees that they cut down there in the first place is a crime in itself. So I suppose I'm trying to do something to reinstate what was there in the past"*.

Aesthetic
Consistent with previous research, amenity reasons are often a strong motivation for landholders to participate in conservation programmes (Ma et al. 2012). Both amenity migrants (Gosnell and Abrams 2009) and commercial landholders considered biodiverse carbon plantings as a way of increasing aesthetic values of their properties. *"It was about visual ...when you look... doing something to lift the quality of the paddocks, a bit of visual amenity I guess, it was a pretty bare rock farm prior to that"* (**Noah and Linda**). Noah and Linda regard themselves as amenity migrants or "blockies" as they are called by farmers in their area. The enhancement of the visual amenity of their property has been one of the major drivers for their participation. *"I don't know how you measure that or say it but it's nice. People like to plant trees. They don't like an open plain"* (**David**). David, a commercial cattle grazer, thinks that beautifying properties is subjective and there is not a specific metric to gauge the result. To enhance the visual quality of their properties, aesthetic benefits encourage landholders to join the biodiverse carbon planting scheme.

CONCLUSION

In this chapter, I applied adoption theory (Pannell et al. 2006) to explore motivations and barriers for landholders to participate in bio-diverse carbon planting at the first two stages of adoption. I identified important factors identified by landholders that assisted them to become aware of biodiverse carbon planting schemes in the awareness phase. Cultural motivation, the roles of formal and informal education, emergent stewardship, participation in Landcare groups and social networks are among the drivers explored in this stage. Experiencing nature as a child (Louv 2010) and social learning from role models and influential figures in landholders' lives were major cultural motivations for them to participate in biodiverse carbon planting to rehabilitate the land. Furthermore, through their participation in a Landcare group or suggestions from a trusted peer within their social networks, landholders became aware of the scheme.

In the non-trial step, I discussed factors such as co-benefits of bio-diverse carbon plantings that have an impact on landholders' decision-making processes. Both biodiversity- and farm-related co-benefits were held to impact on the programme uptake in this stage. Landholders' desire for biodiversity conservation alongside running their property increased their interest in participating. Furthermore, landholders' "eco-centric" stances (Lockwood 1999) towards landscape connectivity and habitat restoration were an influential factor in their participation. The intrinsic value of native flora as a way of conserving the diversity of species and for its wildlife habitat benefits were also influential factors for landholders. Farm-related co-benefits (Jellinek et al. 2013) like salinity and erosion control, pasture and livestock improvement acted as drivers to motivate landholders to participate. These findings could assist policy-makers with improving adaptive governance in carbon farming policies.

REFERENCES

Addison, J. (2013). *Impact of Climate Change on Health and Wellbeing in Remote Australian Communities: A Review of Literature and Scoping of Adaptation Options* (CRC-REP Working Paper CW014). Alice Springs: Ninti One Limited.

Adger, N. W., Arnell, N. W., & Tompkins, E. L. (2005). Successful Adaptation to Climate Change Across Scales. *Global Environmental Change, 15*(2), 77–86. Retrieved from http://linkinghub.elsevier.com/retrieve/pii/S0959378004000901.

Ahnström, J., Höckert, J., Bergeå, H. L., Francis, C. A., Skelton, P., & Hallgren, L. (2008). Farmers and Nature Conservation: What Is Known About Attitudes, Context Factors and Actions Affecting Conservation? *Renewable Agriculture and Food Systems, 24*(1), 38–47. Retrieved from http://www.journals.cambridge.org/abstract_S1742170508002391.

Australian Government Clean Energy Regulator. (2014). *Carbon Farming Initiative*. Retrieved from http://www.cleanenergyregulator.gov.au/Carbon-Farming-Initiative/Pages/default.aspx.

Balnaves, M., & Caputi, P. (2001). *Introduction to Quantitative Research Methods: An Investigative Approach*. London: Sage.

Baumgart-Getz, A., Prokopy, L. S., & Floress, K. (2012). Why Farmers Adopt Best Management Practice in the United States: A Meta-Analysis of the Adoption Literature. *Journal of Environmental Management, 96*(1), 17–25. Retrieved from http://www.ncbi.nlm.nih.gov/pubmed/22208394.

Berry, W. (1995). *Another Turn of the Crank: Essays*. Washington, DC: Counterpoint Press.

Bigsby, H. (2009). Carbon Banking: Creating Flexibility for Forest Owners. *Forest Ecology and Management, 257*(1), 378–383. Retrieved from http://linkinghub.elsevier.com/retrieve/pii/S0378112708006920.

Blackmore, L., & Doole, G. J. (2013). Drivers of Landholder Participation in Tender Programs for Australian Biodiversity Conservation. *Environmental Science & Policy, 33*, 143–153. Retrieved from http://linkinghub.elsevier.com/retrieve/pii/S1462901113001226.

Blaikie, N. (2000). *Designing Social Research: The Logic of Anticipation*. Cambridge: Polity.

Bowen, M. E., McAlpine, C. A., House, A. P. N., & Smith, G. C. (2007). Regrowth Forests on Abandoned Agricultural Land: A Review of Their Habitat Values for Recovering Forest Fauna. *Biological Conservation, 140*(3–4), 273–296. Retrieved from http://linkinghub.elsevier.com/retrieve/pii/S0006320707003308.

Briggs, B. S. V. (2001). Linking Ecological Scales and Institutional Frameworks for Landscape Rehabilitation. *Ecological Management and Restoration, 2*(1), 28–35. Retrieved from http://doi.wiley.com/10.1046/j.1442-8903.2001.00065.x.

Bryman, A. (2004). *Social Research Methods* (2nd ed.). Oxford: Oxford University Press.

Bryman, A., & Burgess, R. G. (1999). Qualitative Research Methodology: A Review. In *Qualitative Research, Vol. 1, Fundamental Issues in Qualitative Research*. London: Sage.

Buijs, A. (2009). Lay People's Images of Nature: Comprehensive Frameworks of Values, Beliefs, and Value Orientations. *Society & Natural Resources,*

22(5), 417–432. Retrieved from http://www.tandfonline.com/doi/abs/10.1080/08941920801901335.

Burton, R. J. F., & Paragahawewa, U. H. (2011). Creating Culturally Sustainable Agri-Environmental Schemes. *Journal of Rural Studies, 27*(1), 95–104. Retrieved from http://linkinghub.elsevier.com/retrieve/pii/S0743016710000720.

Chawla, L. (1999). Life Paths into Effective Environmental Action. *The Journal of Environmental Education, 31*(1), 15–26. Retrieved from http://www.tandfonline.com/doi/abs/10.1080/00958969909598628.

Comerford, E. (2014). Understanding Why Landholders Choose to Participate or Withdraw from Conservation Programs: A Case Study from a Queensland Conservation Auction. *Journal of Environmental Management, 141*, 169–176. Retrieved from http://linkinghub.elsevier.com/retrieve/pii/S0301479714000784.

Commonwealth of Australia. (2013). *Australia's Abatement Task and 2013 Emissions Projections.* Canberra: Commonwealth of Australia.

Compton, E., & Beeton, R. J. S. (Bob). (2012). An Accidental Outcome: Social Capital and Its Implications for Landcare and the "Status Quo". *Journal of Rural Studies, 28*(2), 149–160.

Cooke, B., & Lane, R. (2015). How Do Amenity Migrants Learn to Be Environmental Stewards of Rural Landscapes? *Landscape and Urban Planning, 134*, 43–52. Retrieved from http://linkinghub.elsevier.com/retrieve/pii/S0169204614002382.

Cooper, S., Endacott, R., & Chapman, Y. (2009). Qualitative Research: Specific Designs for Qualitative Research in Emergency Care? *Emergency Medicine Journal, 26*(11), 773–776.

Dohrenwend, B. (1965). Some Effects of Open and Closed Questions on Respondents' Answers. *Human Organization, 24*(2), 175–184. Retrieved from http://sfaa.metapress.com/content/U5838W33858455U3.

Doole, G. J., Blackmore, L., & Schilizzi, S. (2014). Determinants of Cost-Effectiveness in Tender and Offset Programmes for Australian Biodiversity Conservation. *Land Use Policy, 36*, 23–32. Retrieved from http://linkinghub.elsevier.com/retrieve/pii/S0264837713001166.

Ecker, S., Thompson, L., Kancans, R., Stenekes, N., & Mallawaarachchi, T. (2012). *Drivers of Practice Change in Land Management in Australian Agriculture.* ABARES Report to Client Prepared for Sustainable Resource Management Division, Canberra.

Fischer, A. P., & Bliss, J. C. (2008). Behavioral Assumptions of Conservation Policy: Conserving Oak Habitat on Family-Forest Land in the Willamette Valley, Oregon. *Conservation Biology: The Journal of the Society for Conservation Biology, 22*(2), 275–283. Retrieved from http://www.ncbi.nlm.nih.gov/pubmed/18241236.

Gill, N. (2013). Making Country Good: Stewardship and Environmental Change in Central Australian Pastoral Culture. *Transactions of the Institute of British Geographers, 39*(2), 265–277.

Glaser, B., & Strauss, A. L. (1967). *The Discovery of Grounded Theory: Strategies for Qualitative Research.* Chicago: Aldine Publishing Company.

Gosnell, H., & Abrams, J. (2009). Amenity Migration: Diverse Conceptualizations of Drivers, Socioeconomic Dimensions, and Emerging Challenges. *GeoJournal, 76*(4), 303–322. Retrieved from http://link. springer.com/10.1007/s10708-009-9295-4.

Hartley, M. J. (2002). Rationale and Methods for Conserving Biodiversity in Plantation Forests. *Forest Ecology and Management, 155*(1–3), 81–95. Retrieved from http://www.sciencedirect.com/science/article/pii/S0378112701005497.

Holstein, J. A., & Gubrium, J. F. (2003). *Inside Interviewing: New Lenses, New Concerns* (J. A. Holstein & J. F. Gubrium, Eds.). Thousands Oaks: Sage.

Hulvey, K. B., Hobbs, R. J., Standish, R. J., Lindenmayer, D. B., Lach, L., & Perring, M. P. (2013). Benefits of Tree Mixes in Carbon Plantings. *Nature Climate Change, 3*(10), 869–874. Retrieved from http://www.nature.com/doifinder/10.1038/nclimate1862.

Jack, B. K., Kousky, C., & Sims, K. R. E. (2008). Designing Payments for Ecosystem Services: Lessons from Previous Experience with Incentive-Based Mechanisms. *Proceedings of the National Academy of Sciences of the United States of America, 105*(28), 9465–9470.

Jellinek, S., Parris, K. M., Driscoll, D. A., & Dwyer, P. D. (2013). Are Incentive Programs Working? Landowner Attitudes to Ecological Restoration of Agricultural Landscapes. *Journal of Environmental Management, 127,* 69–76. Retrieved from http://linkinghub.elsevier.com/retrieve/pii/S0301479713002788.

Johnson, J. M.-F., Franzluebbers, A. J., Weyers, S. L., & Reicosky, D. C. (2007). Agricultural Opportunities to Mitigate Greenhouse Gas Emissions. *Environmental Pollution, 150*(1), 107–124. Retrieved from http://www.ncbi.nlm.nih.gov/pubmed/17706849.

Kahn, P. H., & Kellert, S. R. (Eds.). (2002). *Children and Nature: Psychological, Sociocultural, and Evolutionary Investigations.* Cambridge, MA: MIT Press.

Kahn, P. H., Severson, R. L., & Ruckert, J. H. (2009). The Human Relation with Nature and Technological Nature. *Current Directions in Psychological Science, 18*(1), 37–42. Retrieved from http://cdp.sagepub.com/lookup/doi/10.1111/j.1467-8721.2009.01602.x.

Kilpatrick, S., & Johns, S. (2003). How Farmers Learn: Different Approaches to Change. *The Journal of Agricultural Education and Extension, 9*(4), 151–164. Retrieved from http://www.tandfonline.com/doi/abs/10.1080/13892240385300231.

Kragt, M. E., Blackmore, L., Capon, T., Robinson, C. J., Torabi, N., & Wilson, K. A. (2014). *What Are the Barriers to Adopting Carbon Farming Practices?* (Working Paper No. 1407). School of Agricultural and Resource Economics, University of Western Australia. Retrieved from http://ageconsearch.umn. edu/handle/195776.

Latchem, R. (2014). Informal Learning and Non-formal Education for Development. *Journal of Learning for Development, 1*(1). Retrieved from https://files.eric.ed.gov/fulltext/EJ1106082.pdf.

Lauber, T. B., Decker, D. J., & Knuth, B. A. (2008). Social Networks and Community-Based Natural Resource Management. *Environmental Management, 42*(4), 677–687. Retrieved from http://www.ncbi.nlm.nih. gov/pubmed/18704565.

Lin, B. B. (2011). Resilience in Agriculture Through Crop Diversification: Adaptive Management for Environmental Change. *BioScience, 61*(3), 183–193. Retrieved from http://bioscience.oxfordjournals.org/cgi/ doi/10.1525/bio.2011.61.3.4.

Lockwood, M. (1999). Humans Valuing Nature: Synthesising Insights from Philosophy, Psychology and Economics. *Environmental Values, 8*(3), 381–401.

Louv, R. (2010). *Last Child in the Woods: Saving Our Children from Nature-Deficit Disorder* (Rev. ed.). London: Atlantics.

Ma, Z., Butler, B. J., Kittredge, D. B., & Catanzaro, P. (2012). Factors Associated with Landowner Involvement in Forest Conservation Programs in the U.S.: Implications for Policy Design and Outreach. *Land Use Policy, 29*(1), 53–61. Retrieved from http://linkinghub.elsevier.com/retrieve/pii/ S0264837711000457.

Mankad, A., Walton, A., & Alexander, K. (2015). Key Dimensions of Public Acceptance for Managed Aquifer Recharge of Urban Stormwater. *Journal of Cleaner Production, 89*, 214–223. Retrieved from http://linkinghub.elsevier. com/retrieve/pii/S0959652614012086.

Martin, P. (1997). The Constitution of Power in Landcare: A Post-structuralist Perspective with Modernist Undertones. In S. Lockie & F. Vanclay (Eds.), *Critical Landcare* (Vol. 5, pp. 45–56). Wagga Wagga: Centre for Rural Social Research (SCU).

Meadows, J., Emtage, N., & Herbohn, J. (2014). Engaging Australian Small-Scale Lifestyle Landowners in Natural Resource Management Programmes—Perceptions, Past Experiences and Policy Implications. *Land Use Policy, 36*, 618–627. Retrieved from http://linkinghub.elsevier.com/retrieve/pii/ S0264837713002111.

Menz, M. H. M., Dixon, K. W., & Hobbs, R. J. (2013). Ecology. Hurdles and Opportunities for Landscape-Scale Restoration. *Science, 339*(6119), 526–527. Retrieved from http://www.ncbi.nlm.nih.gov/pubmed/23372001.

3 UNDERSTANDING STAKEHOLDERS ...   45

Moon, K., & Cocklin, C. (2011). A Landholder-Based Approach to the Design of Private-Land Conservation Programs. *Conservation Biology: The Journal of the Society for Conservation Biology, 25*(3), 493–503. Retrieved from http://www.ncbi.nlm.nih.gov/pubmed/21309851.

Morris, J., Mills, J., & Crawford, I. M. (2000). Promoting Farmer Uptake of Agri-Environment Schemes: The Countryside Stewardship Arable Options Scheme. *Land Use Policy, 17*(3), 241–254. Retrieved from http://linkinghub.elsevier.com/retrieve/pii/S0264837700000211.

Morrison, M., Durante, J., Greig, J., & Ward, J. (2008). *Encouraging Participation in Market Based Instruments and Incentive Programs.* Canberra: Land & Water Australia.

Öhlmér, B. (1998). Understanding Farmers' Decision Making Processes and Improving Managerial Assistance. *Agricultural Economics, 18*(3), 273–290. Retrieved from http://www.sciencedirect.com/science/article/pii/S0169515097000522.

Pahl-Wostl, C., Mostert, E., & Tàbara, D. (2008). The Growing Importance of Social Learning in Water Resources Management and Sustainability Science. *Ecology and Society Society, 13*(1), 24.

Pannell, D. J., Marshall, G. R., Barr, N., Curtis, A., Vanclay, F., & Wilkinson, R. (2006). Understanding and Promoting Adoption of Conservation Practices by Rural Landholders. *Australian Journal of Experimental Agriculture, 46*(11), 1407–1424. Retrieved from http://www.publish.csiro.au/?paper=EA05037.

Rankin, S. (2014). Soggy Biscuit. In R. Archer, R. Scott, A. Pung, J. Hearn, C. Cliff, K. Olsson, ... S. Varga (Eds.), *Griffith Review 44: Cultural Solutions.* Brisbane: Griffith University.

Rochecouste, J.-F., Dargusch, P., Cameron, D., & Smith, C. (2015). An Analysis of the Socio-Economic Factors Influencing the Adoption of Conservation Agriculture as a Climate Change Mitigation Activity in Australian Dryland Grain Production. *Agricultural Systems, 135*, 20–30. Retrieved from http://linkinghub.elsevier.com/retrieve/pii/S0308521X1400170X.

Saunders, D., & Walker, B. (1998). Biodiversity and Agriculture. *Reform* (Spring) (6), 11–16.

Sobels, J., Curtis, A., & Lockie, S. (2001). The Role of Landcare Group Networks in Rural Australia: Exploring the Contribution of Social Capital. *Journal of Rural Studies, 17*(3), 265–276. Retrieved from http://linkinghub.elsevier.com/retrieve/pii/S0743016701000031.

Standish, R. J., & Hulvey, K. B. (2014). Co-benefits of Planting Species Mixes in Carbon Projects. *Ecological Management & Restoration, 15*(1), 26–29. Retrieved from http://doi.wiley.com/10.1111/emr.12084.

Stern, S. (2008). Reconsidering "Crowding Out" of Intrinsic Motivation from Conservation Incentives. *Critical Issues in Environmental Taxation, 24,* 1–20.

Stoneham, G., Chaudhri, V., & Ha, A. (2003). Auctions for Conservation Contracts: An Empirical Examination of Victoria's Bushtender Trial. *The Australian Journal of Agricultural and Resource Economics, 47*(4), 477–500.

Thompson, S. C. G., Barton, M. A., College, P., & Avenue, H. (1994). Ecocentric and Anthropocentric Attitudes Toward the Environment. *Journal of Environmental Psychology, 14,* 149–157.

Torabi, N., Cooke, B., & Bekessy, S. A. (2016). The Role of Social Networks and Trusted Peers in Promoting Biodiverse Carbon Plantings. *Australian Geographer, 47*(2), 139–156. http://doi.org/10.1080/00049182.2016.115 4535.

Trigger, D., Toussaint, Y., & Mulcock, J. (2010). Ecological Restoration in Australia: Environmental Discourses, Landscape Ideals, and the Significance of Human Agency. *Society & Natural Resources, 23*(11), 1060–1074. Retrieved from http://www.tandfonline.com/doi/abs/10.1080/08941920903232902.

Wilson, G. A., & Hart, K. (2001). Farmer Participation in Agri-Environmental Schemes: Towards Conservation-Oriented Thinking? *Sociologia Ruralis, 41*(2), 254–274.

CHAPTER 4

Landholders' Sociocultural Drivers Influencing Decision-Making and Participation in Carbon Farming

Abstract Torabi explores the social and cultural drivers of landholders that impact on their decisions about participating in biodiverse carbon schemes. Factors like uncertainty about political and market elements could influence participations are discussed. Understanding these elements could assist the improvement of adaptive governance systems.

Keywords Adoption theory · Decision process · Participation
Uncertainty

> I am deeply passionate about it, I just keep reading and learning and reading and adopting and doing everything ... every opportunity... I mean it's a varied response. It's not like it's a sort of a sparkling moment where everything changed, it's a continual learning curve. (**Oliver, one of the interviewed landholders**)

INTRODUCTION

This chapter explores the role of landholders' sociocultural drivers in decision-making and participation stages of adoption theory in carbon farming schemes. Exploring these drivers could assist policymakers to engage landholders more effectively while moving to an adaptive governance system.

© The Author(s) 2019
N. Torabi, *Adaptive Governance in Carbon Farming Policies*,
https://doi.org/10.1007/978-3-319-97496-5_4

THIRD STEP: TRIAL EVALUATION (DECISION PHASE) OF BIODIVERSE CARBON PLANTING BY LANDHOLDERS

Following the awareness and non-trial evaluation phases in which landholders appraise their motivations for participation in biodiverse carbon planting, trial evaluation is the third phase. Pannell et al. (2006) refer to this phase as including both decision-making and small-scale trials of the new practice. In this phase, landholders will also gauge whether they have adequate skills to undertake the new practice. The trial evaluation (decision phase) leads landholders to the uptake of the programme (Korhonen et al. 2013). In addition to factors related to landholders' characteristics, external factors (such as the sociopolitical context) also impact on their decisions to participate in biodiverse carbon planting

Decision Process

The impact of human decision-making on the conservation of biodiversity is insufficiently studied (Milner-Gulland 2012). Such studies would assist policymakers to understand the potential outcomes of land management policy (Milner-Gulland 2012) by changing the focus of such schemes from a focus on environmental outcomes exclusively to including consideration of the landholders' decision-making (Cooke et al. 2011).

Decision-making by landholders related to conservation practices on their properties are influenced by the conservation values they assign to their properties and also their confidence in possible positive outcomes (Brain et al. 2014). However, interview results reveal that the decision process has been easier for landholders who have already been involved in similar conservation practices, such as revegetation. They also expressed that they felt certain about what was involved in the process of joining the programme to plant trees on their properties. "*Well it was easy for us to make the decision in that we already understood what was involved; we were already doing that sort of work. It was more for us to use Greenfleet*" (**George**). In the process of decision-making for participation, a few landholders who were living with their older family members had to reach an agreement about biodiverse carbon plantings. As Daisy explained: "*It was like I think I just told mum what we were doing and waited for the... and I still get it... she still comes in, because my mum, you've got to understand is nearly 92, so she still comes and says, I don't know why you did it. Why did you do it?*" The findings of this study support the assertion of Pannell et al. (2006) that decision-making is complex when the property is run by a team of family members.

Time to Invest

Another factor affecting landholders' decisions is the time to invest in planting trees. Lack of time is one of the major barriers for landholders to participate in conservation programmes, especially if such participation requires them to change their property management from status quo (Moon and Cocklin 2011a, b; Moon et al. 2012; Pannell et al. 2006). Non-participants in biodiversity conservation programmes in North Queensland stated that they had less free time to participate in those programmes (Moon et al. 2012). Some landholders like William indicated that they joined Greenfleet as their own attempts at biodiverse plantings had failed and he lacked the time to invest personally in revegetation. *"The main reason was, as I said, I had a go at it myself, and it was very time consuming and it failed"*. Landholders not only consider the investment time but also take into account the opportunity for more successful outcomes within a shorter time frame.

> Like I said before, it was just an opportunity. For us it just suited us: to get those areas all done out there would have taken probably ten or twelve years to have got to that stage and we did it all in... I think we were told about it in April or something, and it was four months from go to whoa. It put a bit of pressure on getting it all done, but the actual planting was no problem because there was a hundred odd scouts going planting trees like mad. (**Mat**)

Landholder surveys also revealed that all of the participants believe that time to invest impacts on their ability to participate in conservation activities on their property.

External Factors

External factors here refer to the context and sociopolitical settings that have an impact on the decision to participate in the programme (Morris et al. 2000). Some of these factors relate directly to the nature of policy and the political situations at the time such schemes are introduced to landholders. Other factors influencing landholders' decisions relate to the nature of national and global carbon markets and the future prospect of these markets. Both categories will be elaborated in the next section.

Uncertainty

Landholders revealed that uncertainties are one of the major barriers for their participation in the regulated market for biodiverse carbon plantings. Uncertainty, in both political and policy contexts, refers to doubt or confusion which might attend the direction or duration of a policy position, its efficacy or the implications of the ideological perspective of the government of the day. In recognition of the very real implications of this uncertainty for agricultural communities, it must also be noted that uncertainty is the inevitable consequence of the urgent environmental threats of climate change and biodiversity loss (Rockström et al. 2009). Key categories of uncertainties identified in the interviews include programme design (administrative burden), political and carbon market-related.

Administrative Burden

Finding and accessing information, especially at the beginning of a new scheme, is time-consuming for landholders. It may be considered as a burden during the decision-making process for landholders to participate if the administration process seems complicated and if they have problems accessing relevant information in an efficient way.

> I think sometimes people don't know where to go to start finding the information. It does seem to be you know there's DPI and now that's merged with DSE, and you know there's all these government departments and it can be hard for people to know who do I ring? And often when you do ring, you get shoved around within the department. Yeah, so it's probably the more places you can approach, the better really, because a lot of people are a bit threatened I think by ringing government departments. (**Daisy**)

The nature of policy instruments in carbon sequestration and biodiversity conservation matters to landholders. Such programmes would be more attractive to them if delivered with clear rights and responsibilities for landholders. The need for a more appealing and straightforward programme was mentioned by several landholders in this research. "*Farmers aren't interested in anything that's too bureaucratic, so it's got to be reasonably straightforward. So I think it does need to be a partnership and a clear partnership with clear responsibilities*" (**George**).

This echoes Lovell (2010, p. 361) who emphasises the need for "balancing bureaucracy with speed and transparency" in carbon market policies.

Jacob also emphasises the role of a personal contact or a familiar face in facilitating landholders' access to information. "*The trouble is that the whole system is so much red tape now. A lot of the old fellas, you don't talk to a person, it's the same person... personal contact – you can get more information in ten minutes than an hour-and-a-half on the computer*".

Sixty-seven per cent of landholders also revealed in the survey that administrative burden is an important factor impacting on their participation in the regulated carbon market.

Political Uncertainty

I have undertaken this study during a time of rapidly changing policy. Interviews were conducted during the 2013 Federal pre-election period when the nation was concerned about the future of climate policy in Australia (Holmes 2014). The uncertainty about the electoral outcome leads to doubt about policy priorities in the future. Kragt et al. (2014) argue that political uncertainty is one of the major barriers for landholders to participate in carbon farming practices. The lack of a robust institutional framework in carbon policy increases uncertainties in political settings (Paiva and Gomes 2014). Landholder interviews reveal that political uncertainty is one of their major concerns for participation in any regulated carbon planting scheme such as the Carbon Farming Initiative (CFI).

> You know the carbon literacy needs to improve dramatically from our politicians before they'll even move to that space... They're a relatively ignorant group of people in this regard. But I think... I'm talking generational shift here. I think it will happen, but I don't think it's going to happen in the next few years... especially if we have a Coalition Government coming in. I don't hold much hope of that. (**Oliver**)

The uncertain political setting is a hurdle for landholders to participate in biodiverse carbon plantings. They considered the possibility of a change of government as a threat to carbon abatement activity. Hence, any further involvement in such schemes would be very uncertain from their point of view. Mat explained his scepticism towards the policy as a result of such uncertainty.

> I'd say I'm sceptical because I think it's… I reckon we've got about a 90% chance we're going to have a change of government, and that will mean there's going to be a change in the carbon farming initiative. So eventually it will settle down into whatever it's going to do, but quite often with these, they take a long time before they become… what would you say?… user friendly.

There is also a general lack of regulatory assurance in private land conservation-related policies (Raymond and Robinson 2013). Changes of government in Australia can impact on conservation-related strategies. The presence of such uncertainty acts as a barrier for landholders to participate in schemes. *"So what you do today, there's no guarantee of what you're going to be able to do in twenty years if you do something for an investment or a planning… so what I've done is just to create weather breaks and corridors for birdlife and animals"* (**Andrew**).

As Andrew reflected, landholders have undertaken voluntary biodiverse carbon plantings to benefit from landscape restoration and farm and productivity-related effects. They assumed if they were tied to a regulated market-based scheme, the lack of regulatory assurance and certainty in the policies would act as a barrier.

Market Uncertainty

In addition to the existing political uncertainty in Australia, uncertainty about carbon markets and the future of the carbon price is another barrier for landholders to take up carbon farming practices (Kragt et al. 2014; Maraseni and Dargusch 2008). Deficiencies in the market for the carbon stem from the failure to set a price on carbon and the lack of a global carbon market; a well-functioning global carbon market develops flexibility and liquidity (Fankhauser and Hepburn 2010). This needs collaborative work at the international level. Some landholders like Luke expressed their concerns about the uncertainty in the carbon market related to the carbon price and the impact of a reasonable price on landholders' participation.

> No one seems to be able to work out what the actual carbon is worth. I mean, if you got a reasonable price for the carbon credits or whatever it

is, you'd probably find farmers would go to more trouble and plant a few more trees, I think. It's a bit uncertain where that's going to go, where that's leading.

Sixty-seven per cent of landholders considered carbon market stability as a significant factor for their future uptake of carbon farming practices.

Permanence Rule

One of the possible barriers to participation in carbon farming mentioned by landholders is the permanence rule. According to the Kyoto Protocol and the legal aspects of carbon sequestration process, planted trees need to stay on properties for 100 years (Bradshaw et al. 2013). Some landholders were not willing to sign a 100-year agreement.

> Greenfleet gave me two contracts to sign, or a contract to sign, which I did, only too pleased to hand over the carbon credits. They've subsequently come back to me and said, 'No, these contracts don't hold any legal rights now. Will you sign another one?' And that's putting a covenant over the land which they've planted. I said no, that wasn't the original agreement. I'll honour my commitment that you've got the carbon credits but there's no way I'm going to put a covenant on the land because there are parts of the land where I could put gypsy caravans on or something like that when and if the resort ever gets going. So there'd be temporary bird hides, temporary accommodation and those things. (**Jacob**)

Contrary to the experience of Jacob, many landholders have undertaken carbon planting for its co-benefits (Paiva and Gomes 2014) and they were satisfied with having a covenant on their property, guaranteeing the existence of co-benefits (e.g. biodiversity, salinity control) over time. For example, William reflects that he thinks legal binding aspects of carbon plantings are necessary. "*So I thought they sort out all the legals as far as covenants on the land go, which I'm a big fan of, because once the trees are in they can't be removed*".

STEP FOUR: ADOPTION OF BIODIVERSE CARBON PLANTING

Pannell et al. (2006) emphasised that adoption of a new land conservation practice is a "continuous process"; it takes time for landholders to incorporate the practice into their existing land management. Landholder interviews revealed that the key factors influencing their involvement in biodiverse carbon planting during the adoption process include land management alteration, native species preferences and risks (fire and feral/pest animals).

Land Management Alteration

The extent of land management alteration depends on the degree of integration of conservation practices with the existing farming or lifestyle business (Pannell et al. 2006). Blackmore and Doole (2013) found that when a new conservation activity is easy for landholders to adopt, they are more keen to undertake the activity themselves rather than joining a conservation scheme. Landholders who have already been undertaking conservation practices integrated the biodiverse carbon plantings into their agricultural or lifestyle property with less effort.

> What we've done is, as we learnt more about both managing native vegetation and managing pasture, we've renovated the pasture, improved the pasture, improved the water supply and the fencing, set up laneways through the property for ease of management so the investments that we've made there have meant that we can run a lot more stock. So although we don't allow any stock to graze the native vegetation, we've actually doubled our stocking rate from the past. (**George**)

George expressed his prospect about the success of adoption of biodiverse carbon planting and considering it as a way of increasing productivity. This is because he regards it as a part of the bigger picture of managing his property. However, to participate in a regulated market-based scheme like CFI, landholders consider transaction costs such as management, monitoring and verification costs in the adoption phase (Cacho et al. 2013). These could be regarded as barriers to participation in such schemes and need to be incorporated into the policy design (McCann et al. 2005).

Transaction costs for landholders to take up a carbon sequestration scheme such as Greenfleet include fencing the revegetated area and controlling weeds and feral animals. Fencing is the major issue for landholders; they need substantial fences to keep the stock away from their new biodiverse carbon plantings. Like many landholders, David expressed his concerns about establishing fences and changing stock management in favour of the planted area on his property. *"Well, you do have to build about the strongest fence on the farm because on the other side where there's a lot of grass it's a temptation for the stock. So it's probably got to be the strongest fence you have really*". Laura also mentioned the elements of management procedures she needed to undertake to integrate the new plantation to her property management: "*Well, only the agreement with them is you know it was basically weed control, pest control, keeping the stock out until the trees got going. I've done all that, and the trees have grown quite nicely, and so now we're just sort of getting ready to, as I said, put the stock back into it*". However, the transaction costs involved seemed like a massive hurdle for a cohort of landholders. As James expressed, to integrate the new revegetation areas with his existing farm management, he had to resolve some of the management alterations, like putting guards around trees. He also expressed some of the issues with electric fencing—disturbance by falling trees in particular.

> Well, there are some downsides. We rely on electric fencing; the property basically has electric fencing so the trees are something of a nuisance in that respect because they quickly fall over the fences and that short-circuits your electric fencing system. I mean I accept the fact that you can have 20% of your farm in trees and not affect your stocking rate but it's a hard country to get established in down there; it's pretty unforgiving. Like the first Greenfleet plantation, they put in 4,000 and I think about 50 grew. So to be fair to Greenfleet, they came and did it again and now they've had a pretty good strike rate, but only because we were prepared to put guards around all the trees they planted and watered them through the summer, which is far more than we probably anticipated with the initial program, but we got them going and we're happy about that. But there's a lot of work involved if someone wants me to put another 40 hectares into the land; there's a lot of work.

Survey result also shows that 67% of landholders think that establishment costs like site preparation and fencing are important factors that they would consider if they were to participate in a regulated carbon trading scheme like CFI.

Native Species Preferences

Australians have historically praised native species of fauna and flora for their productivity and aesthetic benefits, from early birdwatchers to choice of native trees by urban planners (Trigger et al. 2008). The concept of "Australian bush" and landholders' willingness to restore and reconnect with that element of the landscape was clearly stated by interviewees. Gum trees are treated as a flagship species for the bush.

> We would certainly have liked more gums because gum trees are very native to the area, very Australian, if you know what I mean. It's part of life in the bush in Australia I guess, so we were a bit disappointed in that aspect, we would have liked a better diversity of... as I say, more gums and less wattle, but that wasn't to be. (**Noah and Linda**)

Interestingly, the concept of bush is associated with having gum trees on their properties. It reflects on their sense of identity and the complex relationship with the Australian native species (Rogan et al. 2005; Trigger et al. 2010). Participants in this study similarly expressed sentiments towards "bush" conservation and the spiritual value that the bush encompasses (Rogan et al. 2005). It could relate to the cultural and aesthetic preferences of the type of trees planted.

In addition to appreciating native species as a way of connecting to the concept of "Australian bush", the higher survival rate of native trees also matters to landholders. Recognition of the fact that only native species could survive in the harsh environment was observed among landholders. *"The only ones that grow down there are the ones that grow naturally. Anything else is not very successful"* (**James**). Jacob also acknowledges the fact that local seeds were the reason for the biodiverse carbon planting surviving the adverse climate condition on his property. *"It was all sourced from local seed, so it was very easy for it to survive. And surviving... it's a pretty windy block and it's a very wet block and I thought a lot of the trees would have died because they would have root rot but they haven't, they've survived quite well"*.

All landholders in the survey stated that native trees with a diversity of species are the only option that they will choose for any future revegetation activities on their properties. This is because of cultural connectedness to the land and ecological benefit of those species (Trigger et al. 2008).

Risks

Once trees have been established, landholders can be confronted by risks of fire and feral animals associated with revegetation.

Fire Risk

Victoria is one the most fire-prone areas of the world (Fire Services Commissioner Victoria, 2012). Landholders recognise fire as a natural part of the Australian landscape (Halliday et al. 2012). However, the findings of Jellinek et al. (2013) illustrate that non-adopters of revegetation activities were concerned with the increased fire risk on their properties.

> It's no real... it's not worse than if you've got long grass up there. It's quite a temperate climate. It's only really the couple of months of summer that would be an issue. Look, it's an endemic thing in Australian society I suppose, fire risk, and I mean, what do you do? Cut all the trees down and cut down the fire risk? Probably not a great idea is it, because then you've even more risk of fire because the bloody planet's warming up. Yeah. No-one's really mentioned it. When I was on Council we had a few objections to people who had plantations and stuff about the extra fire risk and that, but it's negligible really, and it's a risk... and you manage that risk as best you can. People have to have fire plans and the like, whether it's grass fire or a bushfire. (**William**)

William emphasised that destruction of the natural environment (e.g. cutting trees) contributes to global warming and a higher fire risk as a result. He also mentioned that fire risk in grassy areas is higher than the revegetated area. Laura explains the inherent risk of fire in Australian landscape; however, the presence of thick grass in the revegetated areas increases the fire risk. *"But you know like it doesn't matter whether you've got re-vegetation or not, you're still going to run the risk. It's just that with that re-vegetation there's so much grass in it, it would just flare up so easily"*. The survey results show that only 25% of landholders stated that revegetation on their property will increase the fire risk.

While Oliver believes that the fire risk from revegetation exists, he also believes that the benefits from revegetation outweigh the risks associated with them.

> The fire risk is a bit of a moot point – it can act as a fire corridor, but at the same time, it can also act as a fire break. So one may discount the other, but yeah, there are risks, but there are risks about running a cropping system where you allow standing stubble for too long too. So I mean there are risks with all of them....You make a decision that the risks far outweighed by the benefit, and that's ultimately why we do what we do.

The findings of this study support the assertion of Halliday et al. (2012) about the complex nature of conserving biodiversity and the fire risk associated with such actions.

Wildlife/Feral Animal Risk

The risk of increasing feral animals or unwanted wildlife could be seen as a barrier to participation in revegetation programmes. Indeed, Jellinek et al. (2013) found that some of the landholders were concerned about the increasing abundance of pest animals in the revegetated and remnant vegetation areas. Trees act as a suitable habitat for unwanted wildlife like foxes, kangaroos and rabbits. *"Oh, there's wildlife risk, there's vermin control – if you don't, you can end up with a harbour for foxes, and we have increasingly this last summer, we've noticed in a lot of the bigger plantations there's a lot more kangaroos coming out of them than there were before in the open country"* (**Oliver**). Andrew also considers the increased number of kangaroos as a risk to his property management. In addition, he raises the issue that neighbours could be affected as a result of that change in his property.

> Oh, we've got lots and lots of kangaroos, which we will have to cut their numbers back because they destroy fences. That's the only thing. I mean I don't mind the kangaroos grazing, but destroying fencing is a big problem, a big problem. And of course we have a few foxes, which doesn't cause me any trouble, but my neighbour runs sheep and he has a bit of trouble with when his lambs are happening, the foxes get there, and the odd rabbit.

Jacob appreciates the fact that biodiverse carbon plantings have provided a suitable habitat for wildlife, but he thinks that it has also become a habitat for feral animals. *"It's turned into a very good habitat for deer and wombat, unfortunately, foxes, kangaroos and these things"*.

However, the survey result reveals that only 19% of the participants think that the biodiverse carbon plantings on their properties have increased the risk of rabbits and pest animals. This view towards the pest animals in revegetated areas was expressed much more strongly by participants in the previous studies (Jellinek et al. 2013).

CONCLUSION

I presented two different groups of factors that impact on landholders' uptake of the programme: landholder-related and external factors. Landholders' time to invest and the individual or family setting had a profound impact on their decisions to participate (Reimer et al. 2011). External factors comprised of various types of uncertainties related to the carbon market and political environment. Another influential barrier identified by landholders was the administrative burden of carbon farming programmes. There is also a general lack of regulatory assurance and scepticism of private land conservation policies that governments offer (Reimer et al. 2011). Landholders appreciated the type of biodiverse carbon planting programme they were already involved in as it seemed straightforward and more appealing than the regulated carbon trading schemes (less uncertainty involved).

My analysis of the adoption stage focused mainly on factors related to land management alteration that landholders faced while adopting new land conservation practices. Risks related to planting trees on properties, such as pest animals and fire (Jellinek et al. 2013), were documented in previous studies as barriers to participation in private land conservation. In this study, landholders recognised such risks but did not consider those factors as barriers to their participation. Fire was considered an "endemic thing in Australian society" [William] to landholders. Their choice of native species and recognition that those species are the only ones that will survive in the area were also documented at this stage. These findings could help the programme design and public understanding elements of adaptive governance model.

REFERENCES

Blackmore, L., & Doole, G. J. (2013). Drivers of Landholder Participation in Tender Programs for Australian Biodiversity Conservation. *Environmental Science & Policy, 33*, 143–153. Retrieved from http://linkinghub.elsevier.com/retrieve/pii/S1462901113001226.

Bradshaw, C. J. A., et al. (2013). Brave New Green World—Consequences of a Carbon Economy for the Conservation of Australian Biodiversity. *Biological Conservation, 161,* 71–90.

Brain, R. G., Hostetler, M. E., & Irani, T. A. (2014). Why Do Cattle Ranchers Participate in Conservation Easement Agreements? Key Motivators in Decision Making. *Agroecology and Sustainable Food Systems, 38*(3), 299–316.

Retrieved from http://www.tandfonline.com/doi/abs/10.1080/21683565. 2013.819479.

Cacho, O. J., Lipper, L., & Moss, J. (2013). Transaction Costs of Carbon Offset Projects: A Comparative Study. *Ecological Economics, 88,* 232–243. Retrieved from http://linkinghub.elsevier.com/retrieve/pii/S0921800912004910.

Cooke, B., Langford, W. T., Gordon, A., & Bekessy, S. (2011). Social Context and the Role of Collaborative Policy Making for Private Land Conservation. *Journal of Environmental Planning and Management, 1–17.* Retrieved from http://dx.doi.org/10.1080/09640568.2011.608549.

Fankhauser, S., & Hepburn, C. (2010). Designing Carbon Markets, Part II: Carbon Markets in Space. *Energy Policy, 38*(8), 4381–4387. Retrieved from http://linkinghub.elsevier.com/retrieve/pii/S0301421510002569.

Halliday, L. G., Castley, J. G., Fitzsimons, J. A., Tran, C., & Warnken, J. (2012). Fire Management on Private Conservation Lands: Knowledge, Perceptions and Actions of Landholders in Eastern Australia. *International Journal of Wildland Fire, 21*(3), 197–209. Retrieved from http://dx.doi.org/10.1071/WF10148.

Holmes, B. (2014). *Federal Election 2013: Issues, Dynamics, Outcomes.* Retrieved from http://www.aph.gov.au/About_Parliament/Parliamentary_Departments/Parliamentary_Library/pubs/rp/rp1314/FedElection2013.

Jellinek, S., Parris, K. M., Driscoll, D. A., & Dwyer, P. D. (2013). Are Incentive Programs Working? Landowner Attitudes to Ecological Restoration of Agricultural Landscapes. *Journal of Environmental Management, 127,* 69–76. Retrieved from http://linkinghub.elsevier.com/retrieve/pii/S0301479713002788.

Korhonen, K., Hujala, T., & Kurttila, M. (2013). Diffusion of Voluntary Protection Among Family Forest Owners: Decision Process and Success Factors. *Forest Policy and Economics, 26,* 82–90. Retrieved from http://linkinghub.elsevier.com/retrieve/pii/S1389934112001979.

Kragt, M. E., Blackmore, L., Capon, T., Robinson, C. J., Torabi, N., & Wilson, K. A. (2014). *What Are the Barriers to Adopting Carbon Farming Practices?* (Working Paper 1407). School of Agricultural and Resource Economics, University of Western Australia. Retrieved from http://ageconsearch.umn.edu/handle/195776.

Lovell, H. C. (2010). Governing the Carbon Offset Market. *Wiley Interdisciplinary Reviews: Climate Change, 1*(3), 353–362. Retrieved from http://dx.doi.org/10.1002/wcc.43.

Maraseni, T. N., & Dargusch, P. (2008). Expanding Woodland Regeneration on Marginal Southern Queensland Pastures Using Market-Based Instruments: A Landowners' Perspective. *Australian Journal of Environmental Management, 15,* 112–114.

McCann, L., Colby, B., Easter, K. W., Kasterine, A., & Kuperan, K. V. (2005). Transaction Cost Measurement for Evaluating Environmental Policies. *Ecological Economics, 52*(4), 527–542. Retrieved from http://linkinghub.elsevier.com/retrieve/pii/S0921800904003568.

Milner-Gulland, E. J. (2012). Interactions Between Human Behaviour and Ecological Systems. *Philosophical Transactions of the Royal Society B: Biological Sciences, 367*(1586), 270–278. Retrieved from http://rstb.royalsocietypublishing.org/content/367/1586/270.abstract.

Moon, K., & Cocklin, C. (2011a). A Landholder-Based Approach to the Design of Private-Land Conservation Programs. *Conservation Biology: The Journal of the Society for Conservation Biology, 25*(3), 493–503. Retrieved from http://www.ncbi.nlm.nih.gov/pubmed/21309851.

Moon, K., & Cocklin, C. (2011b). Participation in Biodiversity Conservation: Motivations and Barriers of Australian Landholders. *Journal of Rural Studies, 27*(3), 331–342. Retrieved from http://linkinghub.elsevier.com/retrieve/pii/S0743016711000258.

Moon, K., Marshall, N., & Cocklin, C. (2012). Personal Circumstances and Social Characteristics as Determinants of Landholder Participation in Biodiversity Conservation Programs. *Journal of Environmental Management, 113*, 292–300. Retrieved from http://www.ncbi.nlm.nih.gov/pubmed/23064247.

Morris, J., Mills, J., & Crawford, I. M. (2000). Promoting Farmer Uptake of Agri-Environment Schemes: The Countryside Stewardship Arable Options Scheme. *Land Use Policy, 17*(3), 241–254. Retrieved from http://linkinghub.elsevier.com/retrieve/pii/S0264837700000211.

Paiva, D. S., & Gomes, G. A. M. de M. (2014). Voluntary Carbon Market and Its Contributions to Sustainable Development: Analysis of the Monte Pascoal—Pau Brazil Ecological Corridor. *International Journal of Innovation and Sustainable Development, 8*(1), 1–16.

Pannell, D. J., Marshall, G. R., Barr, N., Curtis, A., Vanclay, F., & Wilkinson, R. (2006). Understanding and Promoting Adoption of Conservation Practices by Rural Landholders. *Australian Journal of Experimental Agriculture, 46*(11), 1407–1424. Retrieved from http://www.publish.csiro.au/?paper=EA05037.

Raymond, C. M., & Robinson, G. M. (2013). Factors Affecting Rural Landholders' Adaptation to Climate Change: Insights from Formal Institutions and Communities of Practice. *Global Environmental Change, 23*(1), 103–114. Retrieved from http://linkinghub.elsevier.com/retrieve/pii/S0959378012001355.

Reimer, A. P., Thompson, A. W., & Prokopy, L. S. (2011). The Multidimensional Nature of Environmental Attitudes Among Farmers in Indiana: Implications for Conservation Adoption. *Agriculture and Human*

Values, 29(1), 29–40. Retrieved from http://www.springerlink.com/index/10.1007/s10460-011-9308-z.

Rockström, J., et al. (2009). Planetary Boundaries: Exploring the Safe Operating Space for Humanity. *Ecology and Society, 14*(2), 32.

Rogan, R., O'Connor, M., & Horwitz, P. (2005). Nowhere to Hide: Awareness and Perceptions of Environmental Change, and Their Influence on Relationships with Place. *Journal of Environmental Psychology, 25*(2), 147–158. Retrieved from http://linkinghub.elsevier.com/retrieve/pii/S0272494405000241.

Trigger, D., Mulcock, J., Gaynor, A., & Toussaint, Y. (2008). Ecological Restoration, Cultural Preferences and the Negotiation of "Nativeness" in Australia. *Geoforum, 39*(3), 1273–1283.

Trigger, D., Toussaint, Y., & Mulcock, J. (2010). Ecological Restoration in Australia: Environmental Discourses, Landscape Ideals, and the Significance of Human Agency. *Society & Natural Resources, 23*(11), 1060–1074. Retrieved from http://www.tandfonline.com/doi/abs/10.1080/08941920903232902.

CHAPTER 5

Understanding Stakeholders: Post-adoption in Carbon Farming

Abstract This chapter explores the experiences of landholders about the established trees on their properties. This is the final step of adoption theory. Hence, understanding these factors could provide an opportunity for a long-term success of carbon farming schemes. Torabi discusses how this fits into the elements of adaptive governance.

Keywords Post-adoption · Benefits · Fauna and flora · Resilience

Introduction

This chapter explores the role of landholders' sociocultural drivers in post-adoption of carbon farming schemes. Exploring these drivers is important as moving towards adaptive governance requires understanding multiple stakeholders' motivations to develop systems that would engage different actors in the process of moving from the status quo.

Step Five: Post-adoption of Biodiverse Carbon Planting

Post-adoption in this study refers to the time when trees are established and landholders start experiencing the benefits. This section focuses on the potential benefits which landholders consider regarding their participation in biodiverse carbon plantings, including socio-ecological resilience and variation in property values. Pannell et al. (2006) argue that

N. Torabi, *Adaptive Governance in Carbon Farming Policies*,
https://doi.org/10.1007/978-3-319-97496-5_5

the final stages after the adoption of the practice could include "modification and non-adoption". Non-adoption was not applicable in the case of biodiverse carbon planting and the participants in this study. This is because once trees are planted they stay on the properties for 100 years as stated in the signed binding agreement between landholders and the offset provider.

Experiencing the Benefits

Landholders undertake biodiverse carbon plantings with a range of different motivations. Some of the benefits experienced in the post-adoption stage are additional to the motivations they had to participate. Some, like Daisy, have participated in biodiverse carbon planting to enhance the biodiversity conservation on their properties but in the post-adoption phase, they could experience other benefits as well. "*Yeah, so it will definitely benefit our land because it was just… yeah the soil was just sort of sitting there getting tilled every few years and probably not getting much chance to sort of recover every year*". Anna also reflected on the fact that she started with the motivation of improving biodiversity conservation but in the post-adoption phase, she has been experiencing the satisfaction of being part of a program in addition to the aesthetic gains. "*Oh, just the amenity, and more birds and the personal satisfaction that I've been part of something. I think that's … I think it's important*".

In addition, some landholders were motivated to participate by increasing the visual amenity on their property. Their reflection on the benefits they have been experiencing was the same as their motivation.

> It no longer looks just like a rock farm. There is a fair bit of visual amenity to it.
>
> You can see why they bring people to come and have a look at it. And they keep telling us it's one of the most successful plantings they've ever done. And you can believe it. And you haven't actually seen it when there was nothing there, but the difference is quite stark. (**Noah and Linda**)

Other landholders who took into account other farm-related benefits (salinity and erosion control, pasture improvement) also mentioned experiencing those benefits in the post-adoption phase. George summarised the farm-related benefits of trees. "*The benefit from the bush is from an agricultural point of view of shelter for stock, and particularly when we were calving cows down, virtually every paddock had good shelter and in winter when it is wet and windy and it's quite cruel for animals*".

Furthermore, the long-term nature of returns from participation in carbon farming schemes requires more recognition (Mitchell et al. 2012). Some landholders also mentioned the need to recognise that benefits from biodiverse carbon plantings are long term. The benefits landholders will experience would not be limited to visual amenities and have both biodiversity and productivity-related advantages on their property.

> But you know like he [a neighbour] would love to do things like that too, and there are people around who, I think if they realised the benefits of doing it, not just aesthetic reasons, but the actual benefits to their pastures and stock feed. You know, it all doesn't happen overnight, and you know I'm just going to start seeing some benefits now, but I look back and I think well it's been worth it, even if it's only just for the wildlife, it's been worth it for me. (**Laura**)

Communicating such understanding among landholders could have an impact on the success of program diffusion.

Changes in Property Value

While some previous studies have argued that planting trees on a property could add to its financial value (Polyakov et al. 2015), some landholders expressed that biodiverse carbon plantings have not added financial value to their properties. *"Not really. I think aesthetically they do and for wind protection they do but there's no economic value"*. As Jacob expressed, the visual amenity and farm productivity are the main benefits of those trees on his property. There is also a great concern about the future buyers of their properties among landholders. Many expressed that in a farming society, traditional farmers would prefer to have fewer trees and more agricultural land. Whereas, if they were to sell their property to a lifestyle landholder, those trees would be considered an added value to the property. In addition, some landholders like William reflected that the increased values they see from the established trees are quite different from the monetary value. Those values they assign to the revegetated areas relate to amenity and broader landscape benefits in the face of climate change.

If you talk to some of my cousins and blokes like [...], they say, why are you planting the property? It's useless now, you can't run stock on it, and you're just devaluing it, but you've got to look at the bigger picture I suppose. You don't know when legislation will change, and the climate the way it is, how valuable these particularly parcels will become into the future. **So maybe now it's probably not making the property any more valuable, but I doubt whether it's making it any less, and my argument would be it looks better as well. So it depends what you put value on.** So there's different ways of looking at value and value to the community and the climate and the fauna around the area.... But from a monetary point of view, I don't think it is. I can't see it really changing the value of the property much, it just changes the use really... you're farming trees instead of farming animals.

The findings in the post-adoption phase support those mentioned in other stages that landholders did not hold purely utilitarian stances towards participation in biodiverse carbon plantings. It also reveals how this program may have influenced the way landholders see themselves and their role as "farming trees".

Changes in the Fauna and Flora

Biodiversity conservation, habitat and wildlife restoration were among the motivations that drove landholders' participation. In the post-adoption phase, landholders mentioned the changes they have noticed in the abundance of both fauna and flora species on their properties, with an emphasis on changes to bird species.

Monitoring of birds, so we've got a fair idea of what we have there, and I think the ... certainly, the biomass of wildlife has increased with the increase in habitat, and it's hard to know whether we're more tuned in, but I think... I'd like to say we have more or greater diversity. We haven't got any evidence of greater diversity, but there's certainly more individuals of particularly bird species. So the health of the environment, the habitat has certainly improved enormously over the years. (**George**)

Apart from the growth in wildlife abundance, the improved habitat on George's property has been a substantial benefit in the post-adoption phase. Andrew also claims the increased number of birds on his property, noticing the bird species which had not existed on his property prior to the plantation. This is as a result of improved habitat on his property.

We've planted all the trees, it's 45 species... more than 45 species of birds here now. Birds turn up all the time that I've never seen before because we've got all the gardens here and the nectars and the gum trees and the shelterbelts, and it just goes on. You know we've hawks and eagles and there's dozens of little wrens. And we never had a wren on the place, and I think we've got hundreds of little wrens around now. They come in to feed. I put out feed for them almost every day too.

This finding is aligned with the previous research on the role of habitat and revegetation structure in the increase of the population size and diversity of bird species (Bowen et al. 2007; Martin et al. 2006).

Furthermore, the increase in the rate of natural regrowth occurred as a result of existing biodiverse carbon planting. This is because of the change in the property management, especially in the fenced areas for those trees (e.g. stopping grazing). Steve noticed the increase in both woody and grassy habitat elements.

Together with what's been planted, there's a lot of natural stuff coming back, you know... where there were existing trees that never got burnt down, obviously they throw their seeds out and it's flourishing up, yeah mostly lots of trees and native grasslands. It's native grasslands because it's so steep you can't work there, you've got to walk around. It's just unbelievable.

This provides a balance between ecological and economic benefits of planted trees (MacLeod and McIvor 2006) and has an impact on land rehabilitation and biodiversity conservation on their properties. In addition, Possingham et al. (2015) argue that land restoration (e.g. revegetation) is more cost-effective than protecting habitat.

However, the lack of a fauna and flora survey to provide evidence for the biodiversity benefits was among landholders concerns in the post-adoption phase. "*I mean the one thing being a bird-watcher that I probably wish I had done a little bit more was do some surveys on my own land of the birds to track those changes*". As Daisy explained, undertaking surveys to set the baseline prior to the plantations and in the post-adoption phase to demonstrate the biodiversity gains of carbon plantings could assist stakeholders. Landholders would then have solid evidence of benefits on their properties and in the broader landscape. It could also provide more opportunities for informal learning through increased involvement in the program (Couvet and Prevot 2015). Furthermore, scheme administrators would have tangible landscape outcomes to communicate to landholders in the awareness phase. To achieve the mentioned objectives, some biodiversity monitoring could be incorporated into the regular monitoring of carbon on the properties, followed by some set standards to showcase the conservation outcomes of biodiverse carbon plantings.

Resilience

Resilience has been widely referred to as the ability of socio-ecological systems to recover after any fluctuations occurring in that system (Cosens 2013; Holling 1973). Given that Victoria has experienced many recent extreme events in the face of climate change (e.g. drought and fire), it is important to consider private land conservation practices as one of the means to increase the socio-ecological resilience in agricultural landscapes (Lin 2011; Tang et al. 2012). The resilience benefits of the biodiverse carbon planting could include diversifying income (in the case of regulated markets only) and improvements in ecological resilience.

> Now historically, if you look at things like the Federation Drought, which I think went for four or five years back in '91. If that happened again today, and it will, a huge percentage of farmers will just go broke, you know they'll walk off the properties, and that's their income gone as well. You know there's a very important social component – suicide rates go up under those times of high stress, and that's where you know you can argue that plants in the landscape are going to give some level of resilience and even where… and it's not desirable ecologically, but farmers have put cattle into bush areas and it's saved them. (**George**)

George explained the economic resilience those planted trees provide for the farm when extreme events happen is a means to assist farmers to survive.

Some landholders highlight the importance of scale in considering resilience in the socio-ecological system.

> Oh we're not planting enough acres to affect the weather, I don't think, so it's not really going to have any influence on whether drought happens or not. No, I think you potentially can increase your… Well you don't affect your stocking rate because the more protection you give your stock the better, so there's a benefit there and if you really want to set up these

biodiversity connections through the area and involve significant acreages of land they're going to have to... somebody's going to have to put in a lot of money. (**James**)

As James mentioned, "scale" is an important factor to consider in undertaking biodiverse carbon plantings to achieve a meaningful outcome across a landscape in terms of both increasing resilience and enhancing biodiversity benefits. Recognition of the complexity of socio-ecological systems urges a move from an "optimisation" management method to a more adaptive way of managing both ecological and social benefits to achieve resilience (Cosens 2013). To achieve the scale of biodiverse carbon plantation which affects resilience, "adaptive governance" in multiple scales across landscape seems crucial (Cosens 2013).

Adoption as a Continuous Process

In addition to recognising the long-term benefits of biodiverse carbon plantations, landholders contemplate the fact that conservation activities continue even in the post-adoption stage. This means landholders possibly undertake more conservation (revegetation) activities and do not see adoption as an endpoint and "review and modification" of land management occurs continuously (Pannell et al. 2006).

> We've been planting corridors from already established areas so they link up so the wildlife can move quite happily all over the property. And each paddock, whether it is a paddock of trees or a paddock of grass, is treated as an individual paddock and we try and manage them as such. So it's just an ongoing management program. There'll be more trees go in, we'll plant out 1,500 to 2,000 trees every year and we've got a few bits and pieces, bits of creeks to finish off. You know we've changed... some of the original plantings we did, some of the creeks we're changing the fencelines of those creeks because they don't suit either where the trees are or where the paddock is, you know. We just have to work... **it's a continual adaption of things**.

As Mat explained, even when the formal adoption stage is over, landholders can alter their property management to fit the trees they have planted. This echoes the findings of Pannell et al. (2006) about the review and modification stage. In the adoption process, there is a continuous revision of land management and conservation activities.

Furthermore, landholders seek other sources of funding to undertake more biodiverse carbon plantings to connect to the existing works and enhance the biodiversity and landscape connectivity benefits on their properties.

> I think we definitely will. Some of the tops of hills… do some more connectivity stuff, re-fence some more of the remnant vegetation. Actually, we've got a plan to do some of that this year, so yeah we definitely will… and some of that will be Biodiversity Fund… not Biodiversity Fund… biodiversity market-funded, because we will do some large old trees. You know the biodiversity market? Yeah, you know how that works, yeah, so we'll do some large trees on this place.

As John reflected, the conservation activity on landholders' properties is a continuous practice (Pannell et al. 2006).

Conclusion

Landholders' perspective in the journey towards adaptive governance is an essential part of any new governance model. This understanding could inform the public learning, representation and process design stages as discussed by Scholz and Stifte (2005) as essential elements and challenges to be considered for achieving adaptive governance. Landholders are change agents in any introduced conservation policy. They require to uptake the new scheme and incorporate it into their primary land use. Hence, this chapter focuses on their viewpoint and unpacking their experiences after their participation.

In this chapter, I applied adoption theory (Pannell et al. 2006) to explore landholders' experience with biodiverse carbon planting at the post-adoption stage. The post-adoption phase focused on how landholders experience benefits both on their farms and related to the broader landscape (Polyakov et al. 2015). Property-related benefits included changes to the fauna and flora on their properties after the trees are established. The additional visual amenity that these trees provided (Polyakov et al. 2015) is valued by landholders. Issues such as property price alteration in the post-adoption phase were discussed in the course of this research. Landholders believe that from a traditional farming point of view, biodiverse carbon plantings could be considered to have a negative impact on the monetary value of their property. However, they also believe that the planted trees have a positive impact on the wildlife and property value for a future like-minded buyer. Many studies have focused exclusively on program awareness and adoption phases (Riley 2006; Tarnoczi and Berkes 2009). However, this research looks at various aspects of post-adoption of a private land conservation practice. This could assist the higher rate of program uptake in rural communities, which is necessary for meeting both carbon abatement and biodiversity conservation goals. In addition, it could inform policy design to consider such factors when aiming for the success of a carbon farming scheme. Apart from landscape connectivity that acted as a driver to participate in the scheme in the earlier stages of adoption, resilience and broad landscape benefits were mentioned by landholders.

References

Bowen, M. E., McAlpine, C. A., House, A. P. N., & Smith, G. C. (2007). Regrowth Forests on Abandoned Agricultural Land: A Review of Their Habitat Values for Recovering Forest Fauna. *Biological Conservation, 140*(3–4), 273–296. Retrieved from http://linkinghub.elsevier.com/retrieve/pii/S0006320707003308.

Cosens, B. A. (2013). Legitimacy, Adaptation, and Resilience in Ecosystem Management. *Ecology and Society, 18*(1), 3.

Couvet, D., & Prevot, A.-C. (2015). Citizen-Science Programs: Towards Transformative Biodiversity Governance. *Environmental Development, 13*, 39–45. Retrieved from http://linkinghub.elsevier.com/retrieve/pii/S2211464514000840.

Holling, C. S. (1973). Resilience and Stability of Ecological Systems. *Annual Review of Ecology and Systematics, 4*(1), 1–23. Retrieved from http://dx.doi.org/10.1146/annurev.es.04.110173.000245.

Lin, B. B. (2011). Resilience in Agriculture Through Crop Diversification: Adaptive Management for Environmental Change. *BioScience, 61*(3), 183–193. Retrieved from http://bioscience.oxfordjournals.org/cgi/doi/10.1525/bio.2011.61.3.4.

MacLeod, N. D., & McIvor, J. G. (2006). Reconciling Economic and Ecological Conflicts for Sustained Management of Grazing Lands. *Ecological Economics, 56*(3), 386–401. Retrieved from http://linkinghub.elsevier.com/retrieve/pii/S0921800905004295.

Martin, T. G., McIntyre, S., Catterall, C. P., & Possingham, H. P. (2006). Is Landscape Context Important for Riparian Conservation? Birds in Grassy Woodland. *Biological Conservation, 127*(2), 201–214. Retrieved from http://linkinghub.elsevier.com/retrieve/pii/S0006320705003198.

Mitchell, C. D., Harper, R. J., & Keenan, R. J. (2012). Current Status and Future Prospects for Carbon Forestry in Australia. *Australian Forestry, 75*(3), 200–212. Retrieved from http://www.tandfonline.com/doi/abs/10.1080/0 0049158.2012.10676402.

Pannell, D. J., Marshall, G. R., Barr, N., Curtis, A., Vanclay, F., & Wilkinson, R. (2006). Understanding and Promoting Adoption of Conservation Practices by Rural Landholders. *Australian Journal of Experimental Agriculture, 46*(11), 1407–1424. Retrieved from http://www.publish.csiro. au/?paper=EA05037.

Polyakov, M., Pannell, D. J., Pandit, R., Tapsuwan, S., & Park, G. (2015). Capitalized Amenity Value of Native Vegetation in a Multifunctional Rural Landscape. *American Journal of Agricultural Economics, 97*(1), 299–3147. Retrieved from http://ajae.oxfordjournals.org/cgi/doi/10.1093/ajae/aau053.

Possingham, H. P., Bode, M., & Klein, C. J. (2015). Optimal Conservation Outcomes Require Both Restoration and Protection. *PLoS Biology, 13*(1), e1002052. Retrieved from http://www.ncbi.nlm.nih.gov/pubmed/25625277.

Riley, M. (2006). Reconsidering Conceptualisations of Farm Conservation Activity: The Case of Conserving Hay Meadows. *Journal of Rural Studies, 22*(3), 337–353. Retrieved from http://linkinghub.elsevier.com/retrieve/ pii/S0743016705000926.

Scholz, J. T., & Stifte, B. (2005). *Adaptive Governance and Water Conflict: New Institutions for Collaborative Planning*. Washington, DC: Resources for the Future.

Tang, Q., Bao, Y., He, X., Zhu, B., & Zhang, X. (2012). Farmer's Adaptive Strategies on Land Competition Between Societal Outcomes and Agroecosystem Conservation in the Purple-Soiled Hilly Region, Southwestern China. *Journal of Mountain Science, 9*(1), 77–86. Retrieved from http://link.springer.com/10.1007/s11629-012-2201-4.

Tarnoczi, T. J., & Berkes, F. (2009). Sources of Information for Farmers' Adaptation Practices in Canada's Prairie Agro-Ecosystem. *Climatic Change, 98*(1–2), 299–305. Retrieved from http://www.springerlink.com/ index/10.1007/s10584-009-9762-4.

CHAPTER 6

Similarities and Differences
in Stakeholders' Voices

Abstract This chapter presents the similarities and differences in the opinions of landholders and other stakeholders about participation in biodiverse carbon plantings. Torabi discusses how far the idea of adaptive governance is in practice from what theory explains. This is demonstrated by representing a model of adaptive governance and the empirical evidence for each step of the model.

Keywords Stakeholders' representation · Environmental policy design Theory–practice gap · Change agents

In exploring motivational drivers and key trigger points that influence adoption of biodiverse carbon plantings among landholders, many factors can be distilled. In this chapter, I will elaborate the emergent findings from the interviews with landholders, policy experts and academics. I will present the similarities and differences in the opinions of landholders and other stakeholders about participation in biodiverse carbon plantings.

SCIENCE AND POLICY STAKEHOLDERS' INTERVIEWS

Participants discussed their ideas and concerns about different aspects of biodiverse carbon plantings. This provided me with the opportunity to add the thoughts of policymakers to those of practitioners and landholders in my research. I interviewed 14 stakeholders (eight policy experts and

© The Author(s) 2019
N. Torabi, *Adaptive Governance in Carbon Farming Policies*,
https://doi.org/10.1007/978-3-319-97496-5_6

six academics) from October 2012 to September 2013. I recruited participants through a workshop about "biodiversity offsets" held in November
2011 and through snowball sampling as many of my interviewees introduced and recommended other experts in the field. I chose interviews over
fixed response surveys in an effort to capture rich, nuanced details which are
difficult to elicit in more structured methods. The sample size is justified by
a "sampling to saturation" philosophy whereby interviews continued until
no new themes were emerging. I was careful to select different actors in the
various public and private agencies to capture as much diversity as possible
and not to reach saturation prematurely (Glaser and Strauss 1967). A few
examples of interview questions are listed below:

- What do you think of different carbon planting schemes, both vol-
  untary and paid programmes?
- How do you think private landholders' participation in carbon
  planting could be increased?
- Which additional incentives would help (monetary and
  non-monetary)?
- Which of these incentives are politically feasible both in an
  Australian context and internationally?
- What could be changed in the science-policy-public landscape in
  favour of carbon planting (in both design and implementation)?

Carbon Sequestration or Biodiversity Conservation?

As expressed in Chapter 4, many landholders consider co-benefits of
biodiverse carbon planting as their primary motivation for participation.
These co-benefits are related to biodiversity conservation and on-farm
co-benefits. In terms of biodiversity conservation, the main drivers for
participation included providing habitat for wildlife, connectivity in
the landscape and landscape restoration. On-farm co-benefits included
increases in productivity through windbreaks and shelterbelts, reducing
erosion and salinity control. Landholders not only consider these benefits
within their property boundary but also acknowledge them as benefits
to a broader landscape—salinity and landscape connectivity in particular.
Interestingly, carbon sequestration was not a driver for the participation
of landholders in this study.

Carbon farming practices (both voluntary and regulated) could fail to adequately recruit landholders if they are marketed to landholders solely on the basis of carbon sequestration. To achieve abatement targets in Australia, it will be important to maximise participation rates in carbon sequestration projects. This research demonstrated that highlighting co-benefits may prove to be more appealing to landholders than framing policies around carbon sequestration and climate change. These are not limited to the farm boundaries; the experience of "doing something" for the public good and improving the condition of natural capital assets is also valued. To achieve this good illustration of co-benefits in understandable terms for landholders is needed. I will now consider some of the sociocultural drivers of participation in biodiverse carbon plantings.

Landholders as Change Agents in Environmental Policy Design

The role of landholders as change agents is not often considered in the design of environmental management policies. This is because such policies are generally designed and delivered to tackle ecological problems, and they are outcome-oriented. However, landholders can have a critical influence on the success of conservation programmes. Therefore, insights into best ways to engage landholders in adopting biodiverse carbon planting schemes could assist policymakers to introduce and implement programmes that are more favourable for the targeted landholders. Landholders are more likely to participate in programmes that are designed in collaboration with them and to address their landscape-related ecological issues (e.g. erosion control).

Similarities and Differences in Stakeholders' Voices

Interviews with landholders, policy experts and academics about biodiverse carbon plantings reveal not only some consistent themes, but also some contrasting perspectives. In Chapters 3–5, I examined the perspectives from landholders. I have also interviewed science and policymakers; here, I compare their voices.

Similarities

Barriers to Participation

Uncertainty was a barrier that all stakeholders agreed upon. Carbon market, policy and political "perceived" uncertainties were common themes concerning all stakeholders in regard to the programme uptake by landholders. Factors related to programme design such as complexity of a programme and administrative burden were also identified as key barriers by all stakeholders. These findings have implications for policymakers and responsible organisations that deliver programmes. To achieve the desired socio-ecological outcomes of biodiverse carbon planting programmes, government bodies should further develop inclusive engagement strategies as these play an important role in landholder willingness to participate.

Improvement in the Public-Policy-Science Landscape

All the stakeholders that participated in this research agreed that the public-policy-science landscape requires some shifts in recognition of different and competing imperatives. From the perspective of landholders, more communication and consultation with the target audience of the introduced policy could assist in greater acceptance. Other stakeholders (policymakers and academics) also stated that existing approaches in the communication of scientific findings to policymakers need to be more efficient and locally meaningful. The often-ineffective nature of such conversations is due to tight time frames and different approaches of science and policy communities in dealing with programmes like biodiverse carbon plantings. In addition, a top-down approach to delivering carbon farming schemes is not likely to be favoured by the community. Landholders in different socio-ecological contexts will not respond to policies in similar ways. Recognition of these differences is vital to local participation. Hence, landholder consultation groups should be tailored to the target audiences.

Contrasting Views

Food Security

Contrasting viewpoints were revealed among the different stakeholders when discussing food security. Landholders generally consider marginal and non-productive land as potential space to revegetate. There is a subset of landholders with a holistic approach to balancing agriculture/land management and conservation activities. They expressed recognition that by sacrificing some prime agricultural land for revegetation, they managed to increase their productivity in the longer term. However, some landholders expressed concern about negative comments they had received about altering good cropping paddocks to biodiverse carbon plantings. Other stakeholders expressed a concern for balancing food security and carbon farming. They argued that we need to consider the increasing demand for food production when designing and implementing such conservation schemes.

Climate Change

Policy and academic stakeholders agreed that biodiverse carbon planting is a means to tackle climate change and achieving the carbon abatement target. This justifies the way that related policies are designed and communicated, especially around carbon sequestration benefits as the focus is on the desired outcomes. As discussed in Chapter 4 and earlier in this chapter, the co-benefits of biodiverse carbon plantings are the most appealing factor for landholders to participate in programmes; tackling climate change was not a driver expressed by landholders. As a result, the focus of the communications to increase landholder participation requires refinements and recognition of varied objectives, rather than imposing a priority on "tackling climate change".

Sociocultural Drivers of Landholders

Science and policy stakeholders expressed the complexity of landholders' sociocultural drivers for programme uptake and decision-making for participation in biodiverse carbon plantings. However, these social drivers were not unpacked by academics and policymakers in the course of interviews. Science and policy stakeholders acknowledged the complexity and the existence of these motivations but could not verify them in particular, which highlights the urgency of this research. This is because the terms "social and cultural drivers" are mainly used in literature in very broad terms. In this study, I unpacked some of those elements from the landholder point of view. This study could assist stakeholders in both research and policy realms to understand the drivers in each stage of programme adoption and use them in favour of conservation and greenhouse gas emission reduction objectives. It may be argued there is an integrated capacity building benefit in the inclusive engagement of landholders through recognition of their identified concerns.

FINANCIAL INCENTIVES

Alternative methods of delivering financial incentives to landholders were explored in Torabi and Bekessy (2015). These include status quo (carbon-only payments and aid from the Biodiversity Fund), bundling carbon and biodiversity credits (offering a higher price for biodiverse carbon plantings) and stacking carbon and biodiversity credits on one piece of land (creating the opportunity for double dipping). While one of the participants was in favour of the status quo and believed that governments are already offering adequate incentives to engage landholders in biodiverse carbon planting adoption, all others stated that bundling and stacking credits are better ways of increasing landholders' willingness to participate. Bundling was discussed as the preferred method of incentivising landholders. This is because market- and policy-related constraints (e.g. additionality) are minimised with bundled credits. This is because bundling scenario does not have the additionality barrier, but stacking has to deal with this barrier. However, several improvements were identified for bundled credits to fully realise potential benefits. Robust standards are required to verify and report on the biodiversity benefits of carbon plantings. Furthermore, modelling techniques need to be improved to capture precise amounts of abated carbon from biodiverse plantations. These elements are underpinned by the need to clarify the objective(s) of bundling and stacking policies. These could be done by engaging scientists in the policy design process. Understanding role of scientists and scientific learning is essential in achieving adaptive governance.

Comparing landholders' perspectives on these financial incentives to those of the other stakeholders reveals that financial incentives feature far less in the dialogue. While landholders indicated that financial incentives could cover some transaction costs, social, cultural and environmental drivers primarily influence their participation. Biodiversity and productivity related co-benefits of biodiverse carbon plantings are among the main factors. When discussing different aspects of a programme, one that offers adequate information and flexibility for participation and fits with landholders' existing land management priorities could be most appealing to landholders. This is partly because landholders may not consider themselves as economic agents and think about other benefits within and beyond their property boundaries (e.g. natural capital, public goods).

In addition, the existing financial gain for small-sized property owners is not considerable (in comparison with the transaction costs). As a result, it does not act as the primary factor influencing participation. This is an important factor to consider in the process design of the policies to ensure we are working towards an adaptive governance.

HOW FAR ARE WE FROM THE ADAPTIVE GOVERNANCE IN CARBON FARMING POLICIES?

Table 6.1 presents the gap between theory and empirical findings in achieving adaptive governance in carbon farming, through reviewing the literature about the definition of adaptive governance, its application, challenges in achieving it and interviews with 31 stakeholders (17 landholders and 14 science and policymakers). The model or theory has five elements: representation, programme design, scientific learning, public learning and problem representativeness. Each step refers to changes that require happening to status quo to be able to work towards an adaptive governance in carbon farming. Science, policy and community are different stakeholders that should be engaged and work together towards innovative approaches for challenging the current top-down governance in carbon farming. Empirical findings from landholders, scientists and policymakers are also represented in the table in the form of direct quotes from the interview results. Comparing each element of the model and the stakeholders' opinion reveals the gap between theory and practice in working towards adaptive governance in carbon farming.

In terms of representation to hear all the actors' voices especially landholders in rural areas, the role of extension officers has been elaborated by all the stakeholders involved in this research. Programme design needs more innovative ways to ensure that all actors are engaged and involved in the decision-making process. This is not straightforward as each actor has their own version of performance criteria. Scientific learning for policymaking should be revised to increase the effectiveness of carbon abatement policies. Landholders revealed that they are keen to be part of the public learning process. However, a gap exists in engaging them in the process of achieving adaptive governance. Due to the political uncertainty related to the carbon abatement policies in Australia, it is hard to measure the effectiveness of any introduced policy.

Table 6.1 The gap between adaptive governance's theory and practice as documented through the empirical research

The model (theory)	Landholders' opinion (interview results)	Science and policy stakeholders (interview results)	Gap between theory and practice
Representation: "who is involved" in the new institutions and procedures	"I suppose they used to have a lot of extension officers, but that whole phase of land holder extension sort of died about ten or fifteen years ago. They don't have people going out on farms showing you how to do stuff anymore, and DSE and DPI [local government organisations] are getting smaller". **Paul**	There's clearly information extension kind of stuff that helps. People aren't going to adopt things they don't know about or don't understand. So yeah you know there'd be a role for that, and if it's just the carbon we'd expect the carbon companies to do that. You know it's not something the government necessarily needs to do except possibly they may end up actually deciding to do the opposite to educate farmers to be very careful of some of the carbon cowboys	The role of extension officers in disseminating schemes and assisting landholders were highlighted. This is a gap that was mentioned by all stakeholders in achieving adaptive governance
Process design: "what mechanisms are in place" to ensure that decision-making considers all stakeholders and their needs	I don't think it's quite there yet, and I think that's interesting with the next generation coming through now whether they actually acknowledge that you probably need to have a fair bit of good will in understanding at a policy level before that will happen. I'm not sure that's going to happen in the immediate future, but I think over time it will because people will try to combine the sort of environmental stewardship principles, and some of that broader you know NRM protection stuff with carbon, and I think that an integrated approach to that, particularly when you do it in mainstream agriculture production landscapes, I think would be very exciting	It does raise the question of how well the government will evaluate each of the options and treat them and whether they will have appropriate monitoring in place so that over time; the methods that do work and methods that don't work can be sorted out but in ways that are not adversely affecting the people who were being doing the work but more from of a sort of a learn as you go policy kind of direction, and I think that approach is good to support innovation which you know people think is really important	A gap exists between theory and practice in the process design. Both landholders and science, and policymakers that no mechanisms are in place to engage all stakeholders in the process

(continued)

Table 6.1 (continued)

The model (theory)	Landholders' opinion (interview results)	Science and policy stakeholders (interview results)	Gap between theory and practice
Scientific learning: requires scientists and policymakers to work more closely together in natural resource management	Carbon price needs to happen, and it needs to be a meaningful price, it needs to be something that's actually going to make some decent coin to help the newer technologies and the cleaner technologies come online, because until you put that R&D work into it, you're not going to get it getting into mainstream, or it's going to take a lot longer to get it into mainstream	I think also the incentives that you know scientists and individuals face. Personally, I work in science related to policy, and I find it really difficult to publish a lot of the work that we do. So, we do stuff that's relevant to policy. We do stuff for policy, you know we've done things for the Department of Climate Change related to carbon markets. We've done a whole series of projects and what I'm finding most difficult is then publishing them in scientific journals because it's essentially not scientific enough. It's too broad and policy-focused rather than being very specific and narrow scientific	A gap exists in seeing the value for the work that science and policy do to develop a more robust outcome. Also, gap exists in the research area that could deliver technologies to enable the effectiveness of the carbon policy
Public learning: involves both engagements of public in the decision-making process. It also indicates that communities need to familiarise themselves with the new process and respect the outcomes	"Well certainly in terms of if they find people and asked their opinion, I would tend to think that people would get involved. Farmers in particular, being mostly conservative, don't want to be told". **Patrick** "I'd be willing to listen and see what they're discussing. I'd be happy to have discussions if they've got something to put forward or they want to talk about, yeah". **Leigh**	Communicating the values to farmers in terms of those multiple benefits. I get very frustrated when we go out there and we communicate with a single issue in mind. Now I know we have to do that from a policy perspective because that's the policy driver, but the reality is that farmers never think singularly; they're always thinking multidimensional, whether they admit it or not, that's how they run their businesses. So being able to communicate that to farmers. The complex science doesn't need to be complex. The message is really simple. It is you plant trees, in particular if you plant trees of a whole lot of different species with a whole lot of different levels, functional groups or whatever you want to describe them, that you end up producing a whole range of benefits, not just for yourself but for the farmer down the road a bit and for the catchment and for the state and for the planet' Farmers know they do that; they take great pride in the fact that they think beyond their property borders	Landholders think that they have not been involved in the decision-making process. They also express that they are keen to be part of the conversation. The gap exists in engaging landholders in the decision-making process

(continued)

Table 6.1   (continued)

The model (theory)	Landholders' opinion (interview results)	Science and policy stakeholders (interview results)	Gap between theory and practice
Problem responsiveness: effectiveness of new policies in dealing with the existing natural resource management challenges while being fair and sustainable	"I'd say I'm sceptical because I think it's... I reckon we've got about a 90% chance we're going to have a change of government, and that will mean there's going to be a change in the carbon farming initiative. So eventually it will settle down into whatever it's going to do, but quite often with these, they take a long time before they become user-friendly". **Scott**	CFI, particularly sequestration projects or abatement projects on land, require changing embedded farming practices. Now it's not just about planting trees or giving up land or the opportunity costs in give up land, it's also things like putting GPS systems on trackers. There are changing skills, education that is required. We're talking about turning around ecosystem processes or we're talking about turning around embedded farming practices. That's changing facilities, equipment, distribution models. It's not a get rich quick scheme. So unless we have long-term political stability and policy settings that allow for investment to occur it's not going to happen	In practice, due to a lack of policy and political certainty, it is hard to evaluate how effective any new policy is

REFERENCES

Glaser, B., & Strauss, A. L. (1967). *The Discovery of Grounded Theory: Strategies for Qualitative Research.* Chicago: Aldine.

Torabi, N., & Bekessy, S. A. (2015). Bundling and Stacking in Bio-Sequestration Schemes: Opportunities and Risks Identified by Australian Stakeholders. *Ecosystem Services, 15,* 84–92. Available at: http://linkinghub.elsevier.com/retrieve/pii/S2212041615300206. Accessed September 29, 2015.

CHAPTER 7

Conclusion

Abstract This chapter focuses on discussing the findings of this research in a broader literature of adaptive governance. Recommendations have been made for the carbon farming policies to advance towards adaptive governance by developing systems that would engage different actors in the process of moving from the status quo. Torabi also makes suggestions for the programme design for a more sustainable practice in the agricultural landscapes.

Keywords Stakeholder engagement · Innovative approaches · Flexible schemes · Improved communication

This chapter focuses on discussing the findings of this research in a broader literature of adaptive governance. Through findings from the interviews with multiple stakeholders in the carbon farming realm (landholders, scientists and policymakers), this book presents some of the gaps and challenges to achieve adaptive governance. To move towards adaptive governance, I will discuss recommendations related to each group of stakeholders in carbon farming policies.

© The Author(s) 2019
N. Torabi, *Adaptive Governance in Carbon Farming Policies*,
https://doi.org/10.1007/978-3-319-97496-5_7

LANDHOLDERS

As explained by Scholz and Stifte (2005), representation, programme design and public learning are some elements to consider regarding engaging landholders when achieving adaptive governance, ensuring that governance mechanisms are in place to engage landholders effectively in the process of decision-making.

The following suggestions have emerged through interviews with landholders and may assist the engagement of landholders in the policy design and implementation to achieve adaptive governance:

- *Clear communication of outcomes of the policy/programme*
- *Landholder involvement in the process of decision-making, not only businesses or organisations that are responsible for programme delivery (e.g. not-for-profits, local governments)*
- *Improving the science-policy-community stakeholder engagement*
- *Demonstrating the outcomes of a programme related to the landholders' context without the use of scientific jargon*

SCIENTISTS AND POLICYMAKERS

Representation, programme design, science learning and problem responsiveness are elements to consider for achieving adaptive governance (Scholz and Stifte 2005) when focusing on science and policy stakeholders. My interviews with scientists and policymakers indicate that gaps exist between adaptive governance theory and practice in the current carbon farming policies and programme design.

Scientists demand more innovative methods to measure their research outcomes when working towards an adaptive governance in carbon farming policies. Demonstrating outcomes of the research in a less traditional way requires developing benchmarks to recognise scientists' engagement with the policy and community stakeholders—for example, conducting community education workshops and providing evidence-based policy recommendations.

Policymakers also recognise that policy stability and certainty could help them build a more robust market for carbon abatement and hence achieving desired socio-ecological outcomes. Having such policies and schemes also facilitates the success of problem responsiveness as one of the steps to achieve adaptive governance. Engaging scientists and community stakeholders in different phases of design and implementation of carbon farming policies could increase the success of the developed schemes.

The following suggestions emerged through interviewing policymakers and scientists and could help to work towards adaptive governance:

- *Engaging scientists in the process of developing policies/programmes*
- *Developing alternative tools to measure success for scientists*
- *Stabilising the carbon market policy to measure programme effectiveness*

RECOMMENDATIONS

An important finding from this research that is directly relevant to policy development relates to the need for more flexible programme design that can offer different options depending on the needs of landholders. Such policy should extensively study the target landholders and has engaged them at the very early stages of the policymaking process. Hence, the government could design a more flexible carbon planting scheme with feedback options, localised to the ecology and social needs. This programme has scope for change and alteration based on the feedback received from the local reference groups, reflects the ecological context and social requirements. In doing so, The Commonwealth and the States enshrine involvement of landholders in decision-making for a more bottom-up approach to improve the policies. That includes working with communities to develop a programme that best suits their needs and fits their existing land management priorities. This could increase their participation rate and the likelihood of larger scale landscape restoration.

Communications with landholders should be reframed, focusing more on conservation co-benefits. Currently, the message is framed around tackling climate change and achieving carbon abatement goals. Government and non-governmental agencies involved in the biodiverse carbon planting could establish mechanisms to communicate co-benefits with landholders (both conservation and non-conservation ones). To achieve this objective, demonstrating both biodiversity- and productivity-related co-benefits is essential. Research and innovation could assist in developing indicators to show the value of other co-benefits because of the revegetation on the properties. This could ensure that the scheme is developed based on the community priorities instead of political priorities (Crowley and Coffey 2007).

A further recommendation relates to engaging scientists in multiple aspects of policy design. One is to develop better standards to measure biodiversity co-benefits of carbon plantings and carbon benefits of biodiversity plantings. I recommend that the Commonwealth works closely with scientists to develop biodiversity standards to demonstrate the co-benefits of biodiverse carbon planting. This could accelerate the process of establishing markets for bundled ecosystem services credits (carbon and biodiversity credits in particular). Market stability is essential in the success of adaptive governance process.

My research findings suggest that working with multiple stakeholders could be an efficient way of improving participation. Australian state governments could set up biodiverse carbon planting reference groups at the community level. Members should include identified "champions" in the community, the local Landcare representative and local government representatives. These groups could meet regularly to discuss the opportunities and challenges in landholder participation in such schemes in each stage of programme adoption. The group could also be responsible for organising open days and engaging other landholders with the success of early adopters. This could enhance engaging multiple stakeholders in the policy design and implementation to achieve adaptive governance.

FINAL COMMENTS

This research illustrates possible means to enhance the role of landholders as change agents in achieving adaptive governance, with a focus on carbon farming. I sought to identify the sociocultural triggers influencing the success of biodiverse carbon planting schemes and to discover the importance of programme design and attractiveness in its successful uptake. My research findings emphasise the role of social capital (social networks and trusted peers), landholder awareness and the value of co-benefits in increasing participation rates in carbon farming schemes. These findings have important ramifications for the design and implementation of policy. Environmental management policies could invest more in enhancing the multiple aspects of social capital to increase landholder participation in conservation practices. Building capacity in rural communities often involves strengthening existing social capital. This could be achieved by valuing champions in communities and building trust in the landholder–agency relationship. However, we need to keep in mind that "adaptive governance is never the same in two places; it is messy and often develops organically within the context of socio-ecological systems, but it can also be encouraged with an intervention aimed at boosting adaptive capacity" (Chaffin et al. 2014).

Furthermore, the scale of conservation practices (e.g. biodiverse carbon planting) could have an impact on achieving their ecological outcomes and increasing resilience in the face of climate change in rural societies. In addition, "scale" is an important factor to be considered if such programmes and policies aim to increase landholders "sustainability", "allowing landholders agility to continue" (Verchot et al. 2007, p. 911). This means recognising the dynamic aspects of sustainability in agricultural landscapes and providing opportunities for landholders to increase their productivity and conservation capacity by revegetating the marginal land.

Environmental policy design could benefit through more attention to transdisciplinary perspectives to achieve adaptive governance. However, this approach demands researchers, policymakers and end users (communities) work together in a more adaptive manner (Campbell et al. 2015). Moving from the traditional top-down method to an adaptive governance framework could assist all the stakeholders to achieve their objectives in a more sustainable manner. In doing so, these objectives are more likely to shift from outcome-oriented ones to socio-ecologically sustainable ones. Environmental management policy also needs reform to consider socio-ecological outcomes instead of ecological benefits with the change agent's role elaborated. For such policies to be successful, one needs to answer these questions:

- What does the policy aim to achieve in a socio-ecological landscape?
- Who are the change agents or audiences of these policies?
- Which programmes could work for their socio-ecological setting?

Biodiverse carbon planting is predominantly considered a conservation activity by landholders. The activity is considered in the framework of emergent stewardship and participation is principally motivated by the conservation, amenity or productivity co-benefits. This perspective is contrary to that of policymakers and academics interviewed in this research, who tend to consider biodiverse carbon planting a policy mechanism to tackle climate change and achieve greenhouse gas emission abatement targets. This reveals a mismatch in the understanding of the aims and objectives of such policies from various stakeholders' perspectives. Responsible government institutions tend to communicate the climate change-related co-benefits more than other those related to conservation and productivity. However, communication of these different aspects of the policy could influence its uptake among landholders and could potentially have a negative impact on the participation rate given the climate change scepticism in some rural communities (Raymond and Robinson 2013).

Markets for ecosystem services on private land are gaining attention in Australia (Figgis et al. 2015). This provides an opportunity for innovation in the market and finding new approaches to deal with barriers such as "additionality" (Fitzsimons 2015). Such innovations could offer bundled or stacked ecosystem services credits for buyers and sellers (Torabi and Bekessy 2015). Furthermore, given the complex dynamic and trade-offs between different ecosystems services, the target audience (e.g. communities) perception of the multiple benefits (water quality, carbon sequestration or landscape restoration) requires careful consideration by policymakers. In doing so, it is necessary to move from the traditional decision-making approaches to a more adaptive and collaborative method. Adaptive governance considers sociopolitical and ecological contexts and the link between them in decision-making, in contrast to the traditional top-down approach (Folke et al. 2005; Wyborn 2015). It also requires collaborative approaches among stakeholders in science, policy and community at multiple levels (Wyborn 2015). Adaptive governance could be incorporated into the assessment of trade-offs in ecosystem services. It enables the process of linking different social, ecological and organisational elements (Wyborn 2015).

REFERENCES

Campbell, C. A., et al. (2015). Designing Environmental Research for Impact. *The Science of the Total Environment, 534,* 4–13. Available at: http://www.ncbi.nlm.nih.gov/pubmed/25557212. Accessed February 13, 2015.

Chaffin, B. C., Gosnell, H., & Cosens, B. A. (2014). A Decade of Adaptive Governance Scholarship: Synthesis and Future Directions. *Ecology and Society, 19*(3), 56.

Crowley, K., & Coffey, B. (2007). New Governance, Green Planning and Sustainability: Tasmania Together and Growing Victoria Together. *Australian Journal of Public Administration, 66*(1), 23–37.

Figgis, P., et al. (Eds.). (2015). *Valuing Nature: Protected Areas and Ecosystem Services.* Sydney: Australian Committee for IUCN.

Fitzsimons, J. A. (2015). Private Protected Areas in Australia: Current Status and Future Directions. *Nature Conservation, 10,* 1–23. Available at: http://natureconservation.pensoft.net/articles.php?id=4635. Accessed February 3, 2015.

Folke, C., et al. (2005). Adaptive Governance of Social-Ecological Systems. *Annual Review of Environment and Resources* (pp. 441–473). Palo Alto: Annual Reviews.

Raymond, C. M., & Robinson, G. M. (2013). Factors Affecting Rural Landholders' Adaptation to Climate Change: Insights from Formal Institutions and Communities of Practice. *Global Environmental Change, 23*(1), 103–114. Available at: http://linkinghub.elsevier.com/retrieve/pii/S0959378012001355. Accessed November 1, 2013.

Scholz, J. T., & Stifte, B. (2005). *Adaptive Governance and Water Conflict: New Institutions for Collaborative Planning.* Washington, DC: Resources for the Future.

Torabi, N., & Bekessy, S. A. (2015). Bundling and Stacking in Bio-Sequestration Schemes: Opportunities and Risks Identified by Australian Stakeholders. *Ecosystem Services, 15,* 84–92. Available at: http://linkinghub.elsevier.com/retrieve/pii/S2212041615300206. Accessed September 29, 2015.

Verchot, L. V., et al. (2007). Climate Change: Linking Adaptation and Mitigation Through Agroforestry. *Mitigation and Adaptation Strategies for Global Change, 12*(5), 901–918. Available at: http://link.springer.com/10.1007/s11027-007-9105-6. Accessed February 8, 2015.

Wyborn, C. (2015). Co-productive Governance: A Relational Framework for Adaptive Governance. *Global Environmental Change, 30,* 56–67. Available at: http://linkinghub.elsevier.com/retrieve/pii/S0959378014001769. Accessed November 26, 2014.

INDEX

© The Editor(s) (if applicable) and The Author(s) 2019
N. Torabi, *Adaptive Governance in Carbon Farming Policies*,
https://doi.org/10.1007/978-3-319-97496-5